DNA
LOGIC
DESIGN
Computing
with DNA

DNA LOGIC DESIGN

Computing with DNA

Hafiz Md. Hasan Babu

University of Dhaka, Bangladesh

World Scientific

NEW JERSEY · LONDON · SINGAPORE · BEIJING · SHANGHAI · HONG KONG · TAIPEI · CHENNAI · TOKYO

Published by

World Scientific Publishing Co. Pte. Ltd.

5 Toh Tuck Link, Singapore 596224

USA office: 27 Warren Street, Suite 401-402, Hackensack, NJ 07601

UK office: 57 Shelton Street, Covent Garden, London WC2H 9HE

Library of Congress Control Number: 2023057841

British Library Cataloguing-in-Publication Data
A catalogue record for this book is available from the British Library.

DNA LOGIC DESIGN
Computing with DNA

ISBN 978-981-12-8771-8 (hardcover)
ISBN 978-981-12-8772-5 (ebook for institutions)
ISBN 978-981-12-8773-2 (ebook for individuals)

For any available supplementary material, please visit
https://www.worldscientific.com/worldscibooks/10.1142/13715#t=suppl

Desk Editors: Nambirajan Karuppiah/Julio Hong

Typeset by Stallion Press
Email: enquiries@stallionpress.com

*To my respected great parents and also to
my lovely wife, daughter, and son
who made me possible to write this book.*

Preface

A deoxyribose nucleic acid (DNA) computer is a collection of specially selected DNA strands whose combinations will employ biomolecular manipulation to solve computational problems, at the same time exploring natural processes as computational models. A good experience with biology and computer science is required to build algorithms to be executed in DNA computing. DNA computing is of twofold: one is theoretical and the other is practical. Starting from observing the structure and dynamics of DNA, the theoretical research began to propose formal models of DNA computers for performing theoretical operations. The practical side of DNA computing has progressed at a much slower rate, mainly because the laboratory work is very time-consuming and includes several constraints. The significant properties of DNA computing are (i) dense data storage; (ii) massively parallel computation; and (iii) extraordinary energy efficiency. The information or data instead of being stored in binary digits will now be stored in the form of the bases A, T, G, and C. The ability to synthesize short sequences of DNA artificially makes it possible to use these sequences as inputs for algorithms. DNA has characteristics that enable one to mimic traditional logic operations. DNA prefers to be in double-stranded form, while single-stranded DNA naturally migrates towards complementary sequences to form double-stranded complexes. Complementary sequences pair the bases adenine (A) with thymine (T) and cytosine (C) with guanine (G). The book titled *DNA Logic Design: Computing with DNA* starts with the basics of DNA computing. Then the fundamental operations of DNA computing are described. This book describes arithmetic circuits, combinational circuits, sequential circuits, memory devices, programmable logic devices, and nano processors in DNA computing. The heat and speed calculation techniques in DNA computing are also discussed in the last two chapters of this book.

About the Author

Dr. Hafiz. Md. Hasan Babu is currently working as a Professor in the Department of Computer Science and Engineering, University of Dhaka as well as the Dean in the Faculty of Engineering and Technology of the University of Dhaka, Bangladesh. He is also a Senate Member of the same university. In addition, at present, he is a member (part-time) of the Bangladesh Accreditation Council under the Ministry of Education which is appointed by the Hon'ble President of the People's Republic of Bangladesh. Moreover, Dr. Hasan Babu is a Board Member of Central Depository Bangladesh Limited (CDBL), Central Counterparty Bangladesh Limited (CCBL), Bangladesh Institute of Capital Market (BICM), and Bangladesh Academy for Securities Markets (BASM). He was also a Director of the Board of Directors of Bangladesh Submarine Cable Company Limited. Dr. Hasan Babu was the Chairman of the Department of Computer Science and Engineering of the University of Dhaka from February 19, 2003 to February 18, 2006 and Pro-Vice-Chancellor of the National University of Bangladesh from July 12, 2016 to July 11, 2020. He was also a Professor and worked as the founding Chairman of the Department of Robotics and Mechatronics Engineering, University of Dhaka, Bangladesh. In addition, Professor Dr. Hafiz Md. Hasan Babu worked as a World Bank Resident Information Technology Expert in the Supreme Court of Bangladesh from April 2006 to April 2008 under the Supreme Court Project Implementation Committee (SCPIC) led by the Hon'ble Chief Justice of the Supreme Court of Bangladesh and also served as a World Bank Senior ICT Consultant in the Information Technology Department of Janata Bank Limited, Bangladesh from April 2008 to January 2015. Dr. Hasan Babu obtained his Ph.D. degree in Electronics and Computer Science from Japan under the Japanese Government Scholarship and received his M.Sc. degree in Computer Science and Engineering from the Czech Republic under the Czech Government Scholarship. He also received the DAAD Research Fellowship from Germany.

Dr. Hafiz Md. Hasan Babu was awarded Dr. M.O. Ghani Memorial Gold Medal by the Bangladesh Academy of Sciences in 2017 for his excellent research work in the progress of Physical Sciences in Bangladesh. In addition, he was awarded the University Grants Commission of Bangladesh Gold Medal Award 2017 in Mathematics, Statistics, and Computer Science category for his research work on quantum multiplier-accumulator device. He is currently an Associate Editor of the famous research journal titled *IET Computers and Digital Techniques* published by the Institution of Engineering and Technology of the United Kingdom. He was a member of the Prime Minister's ICT Task Force in Bangladesh. Dr. Hasan Babu was also the President of the Bangladesh Computer Society for the session 2017–2020, the largest professional society in Bangladesh. At present, he is the President of the International Internet Society, Bangladesh Chapter.

Professor Dr. Hafiz Md. Hasan Babu published more than a 100 research papers. Among them, three research papers have received the best research paper award in the International Conferences.

In addition, he has published the following six text books by three famous publishers in the United Kingdom (UK) and the United States of America (USA) for graduate and post-graduate students:

1. Hafiz Md. Hasan Babu, *Quantum Computing: A Pathway to Quantum Logic Design*, IOP (Institute of Physics) Publishing, 2nd Edition, 2020, Bristol, UK.
2. Hafiz Md. Hasan Babu, *Reversible and DNA Computing*, Wiley Publishers, 2021, UK.

3. Hafiz Md. Hasan Babu, *VLSI Circuits and Embedded Systems*, CRC Press (A Publication of Taylor & Francis Group), July 2022, USA.
4. Md. Jahangir Alam, Guoqing Hu, Hafiz Md. Hasan Babu, and Huazhong Xu, *Control Engineering Theory and Applications*, CRC Press (A Publication of Taylor & Francis Group), September 2022, USA.
5. Hafiz Md. Hasan Babu, and Huazhong Xu, *Multiple-Valued Computing in Quantum Molecular Biology*, Volume I, CRC Press (A Publication of Taylor & Francis Group), September 2023, USA.
6. Hafiz Md. Hasan Babu, and Huazhong Xu, *Multiple-Valued Computing in Quantum Molecular Biology*, Volume II, CRC Press (A Publication of Taylor & Francis Group), September 2023, USA.

Acknowledgments

I would like to express my sincerest gratitude and special appreciation to the various researchers in the field of DNA computing. The contents in DNA Logic Design: Computing with DNA book have been compiled from a wide variety of research works, where the researchers are pioneer in their respective fields. All the research articles related to the contents are listed at the end of each chapter.

I am grateful to my great parents and dear family members for their endless support. Most of all, I want to thank my lovely wife Mrs. Sitara Roshan, sweet daughter Ms. Fariha Tasnim, and sweet son Md. Tahsin Hasan for their invaluable cooperation to complete this book.

Finally, I am also thankful to all of those, specially to my beloved students Nitish Biswas, Md. Tareq Hasan, and Rownak Borhan Himel who have provided their immense support and important time to finish this book.

Contents

6. DNA Memory Devices   **137**

List of Figures

List of Tables

Acronyms

ALU	Arithmetic logic unit
BCD	Binary coded decimal
CLB	Configurable logic block
CPLD	Complex programmable logic device
CPU	Central processing unit
DNA	Deoxyribonucleic acid
FPGA	Field-programmable gate array
IR	Instruction register
LUT	Look-up table
MUX	Multiplexer
NTI	Negative ternary inverter
PAL	Programmable array logic
PCR	Polymerase chain reaction
PLA	Programmable logic array
PROM	Programmable read-only memory
PTI	Positive ternary inverter
RAM	Random access memory
RF	Radiofrequency
SIPO	Serial-in parallel-out
SISO	Serial-in serial-out
SPLD	Simple programmable logic devices
STI	Standard ternary inverter
XNOR	Exclusive NOR
XOR	Exclusive OR

Introduction

Deoxyribose nucleic acid (DNA) computing performs computations using biological molecules, rather than traditional silicon chips. It is a modern area of science that recognizes biomolecules as fundamental elements of electronic devices. It has connections to chemistry, software engineering, cell genomics, physics, and mathematics. DNA computing carries the promise of cheap, huge accessible data storage, and an exponential increase in computing power and speed. DNA computing is a new branch of computing that replaces traditional electronic computing with DNA, biochemistry, and molecular biology hardware. Although it all began with Len Adleman's demonstration of a computing application in 1994, the field has now extended to include storage technologies, nanoscale imaging modalities, synthetic controllers, and response networks, among other things. DNA computing is a distributed and parallel computing method. It is useful for solving problems like searching, sorting, merging, pattern recognition, image processing, and encryption that demand high-complexity computations and enormous data sets. The book *DNA Logic Design: Computing with DNA* introduces a structural approach of describing the theory, experiments, and applications of DNA computing. Chapter 1 covers the DNA computing concept. It contains the definitions, merits, demerits, challenges, and motivations of DNA computing. Chapter 2 consists of all fundamental operations in DNA computing like DNA NOT operation, DNA OR operation, DNA NOR operation, etc. Chapter 3 starts with a discussion of different types of arithmetic circuits in DNA computing such as DNA half-adder, DNA full-adder, DNA half subtractor,

DNA full subtractor, N-molecular sequence DNA adders, DNA carry-skip adder, DNA multipliers, DNA dividers, etc. Chapter 4 includes different DNA combinational circuits, such as DNA encoder, DNA decoder, DNA multiplexer, DNA demultiplexer, etc. Chapter 5 discusses different types of DNA sequential circuits, such as DNA SR latch, DNA SR flip-flop, DNA D flip-flop, DNA T flip-flop, DNA JK flip-flop, DNA shift register, DNA ripple counter, DNA synchronous counter, etc. Chapter 6 contains the architecture and applications of different types of memory devices, such as random access memory (RAM), read-only memory (ROM), and programmable read-only memory (PROM), cache memory in DNA computing. Chapter 7 includes different types of programmable logic devices such as programmable logic array (PLA), field programmable gate array (FPGA), and complex programmable logic device (CPLD) in DNA computing. In Chapter 8, readers will get knowledge about the designs of DNA nano processors and their components such as DNA RAM, DNA instruction register (IR), DNA program counter (PC), DNA incrementor circuit, DNA decoder, DNA multiplexer, DNA arithmetic logic unit (ALU), DNA accumulator, etc. Chapters 9 and 10 discuss the heat calculation and speed calculation techniques of DNA computing. As a whole, this book is a great resource for DNA computing researchers, students, and academicians.

Chapter 1

DNA Computing

1.1 Introduction

Deoxyribose nucleic acid (DNA) computing, also known as molecular computing, is an emerging branch of computer science. It is a new approach that allows massively parallel computation, which can solve non-deterministic polynomial time-complete (NP-complete) problems in much less time than conventional computers. DNA computing is also well-suited for solving combinatorial problems. It is a beautiful combination of biochemistry, molecular biology, and computer science that can carry the information to make arithmetic and logic operations. Instead of using traditional silicon chips, it uses biological molecules to perform the computation. Figure 1.1 shows DNA sequences that are used in DNA computing.

Leonard Adleman, a US computer scientist at the University of Southern California, is known as the creator of DNA computing. He first introduced DNA computing when he used DNA to solve large and complex mathematical problems in November 1994. He showed an experimental theory to solve a classic puzzle in mathematics with a seven-point Hamiltonian path problem, better known as the Travelling Salesman problem. The simple puzzle is that the salesman must find the shortest path between seven cities. He has to search for a path in such a way that he visits every city exactly once and returns to the original city (Figure 1.2).

Adelman expressed cities as single-strand DNA, and he solved the problem using DNA molecules in a standard reaction tube. Each base connects with a sugar molecule and a phosphate molecule to form

1

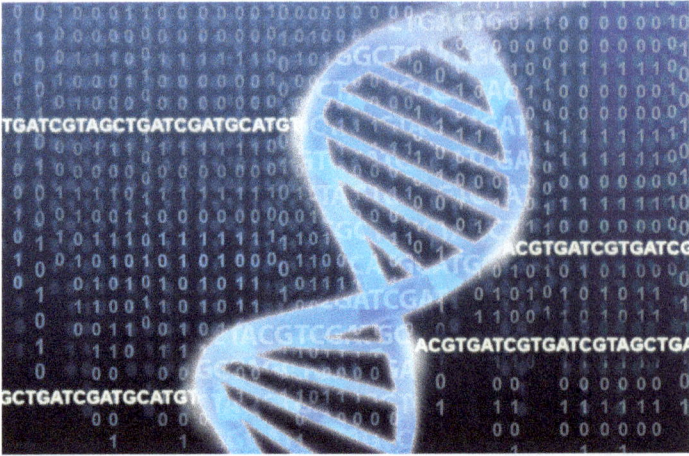

Figure 1.1. DNA sequences.

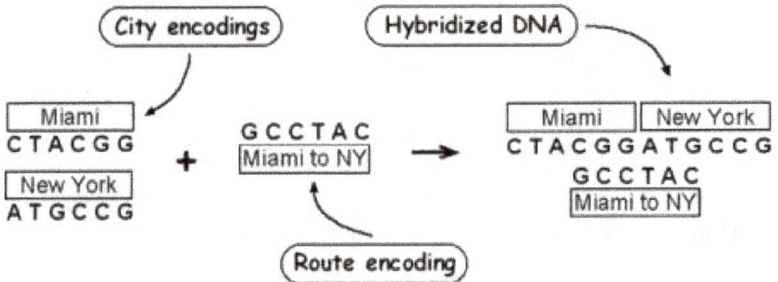

Figure 1.2. Graphical representation of traveling salesman problem.

a nucleotide, of which there are four types: adenine (A), thymine (T), guanine (G), and cytosine (C), which are often abbreviated to letters to represent genetic coding. These four nucleotide bases compose a strand of DNA called a DNA sequence, where the purine A always forms a bond with the pyrimidine T, and the purine G always forms a bond with the pyrimidine C. These are known as Watson–Crick complementary base pairs.

Each DNA strand has two ends and follows a polarity: starting from the 5′-end and ending at the 3′-end. When these bases are mixed in a test tube, they form bonds and pair with each other. Thus all of the possible DNA strand combinations are achieved within a few seconds, representing the answer. The rest of the bases that did not form bonds in this chemical reaction are destroyed. This was the first time DNA computing was applied, and it has now become a successful research field.

DNA computing accomplishes solutions to significant and complex mathematical problems using genetic molecules rather than the traditional silicon substrate. Computations may be executed by using an algorithm, which itself may be defined as a step-by-step list of well-defined instructions that take some inputs and process them. Thus with the right algorithm, any user can produce results from any computation.

In the process of performing computations, one or more techniques for manipulating DNA molecules are used as computational operators for copying, splitting, sorting, or concatenating information: ligation, hybridization, polymerase chain reaction (PCR), gel electrophoresis, and enzymatic reactions.

In the ligation and hybridization steps, all the DNA sequences are added to a test tube using a micropipette. For ligation, these DNA sequences bond with each other to form a DNA strand. For hybridization, the reaction mixture is then heated at 95°C and cooled to 20°C at a rate of 1°C per minute. When a particular DNA sequence with another DNA sequence that is complementary, they will then anneal at complementary base pairs.

PCR performs several functions, such as increasing the amount of a specific DNA molecule in a given mixture. This is done via primer extension by the polymerase. Gel electrophoresis can also be used for sorting DNA strands by their size. Electrophoresis is capable of moving a charged molecule in an electric field. DNA molecules are negatively charged. So, when placed in an electric field, they tend to migrate towards the positive pole. During this migration, smaller molecules move faster through the gel, and thus the molecules can be sorted quickly according to their size.

In DNA computing, information is represented using the letters of the genetic alphabet (A, T, G, and C) rather than the binary alphabet (1, 0). Therefore, all mathematical and significant problems are solved in a DNA computer by encoding a problem using the genetic alphabet A, T, G, and C.

1.2   Merits of DNA Computing

Recently DNA computing has drawn the attention of many researchers because it can solve computationally complex problems. It can perform faster computations than conventional computers because of the nature of its parallelism. Moreover, there are many advantages of DNA computing over silicon-based computing. These advantages are described as follows:

Parallelism: The speed of any computer is determined by two factors: the first one is how many parallel processes it has, and the second one is how many steps each one can perform per unit of time. The most exciting thing is that a single test tube of DNA can contain trillions of strands, and all these strands simultaneously compute biological operations via parallelism.

High information density: DNA can store colossal amounts of information in a small space compared to silicon. For example, one DNA sequence can be easily stored in approximately one cubic nanometer. In other terms, DNA has an estimated storage capacity of 490 exabytes per gram.

Low power consumption: DNA computers can perform 2×10^{19} irreversible operations per joule, whereas supercomputers can execute a maximum of 10^9 operations per joule. DNA computers need power only to prevent DNA from denaturation. In light of predicted energy shortages, this would be a key advantage.

Best-suited for combinatorial problems: DNA computing mainly focuses on solving NP computation and other complicated computational problems. Moreover, it is suitable for solving complex combinatorial problems like the traveling salesman problem, whereas the same would be unimaginable using conventional computers.

Faster speed: As conventional computers are silicon-based, they have a limited rate and cannot be miniaturized beyond a certain point. DNA computers enjoy faster speeds as they already exist beyond at a size much smaller than this point.

No harmful materials: There are no harmful materials in DNA computers. They do not generate any pollution, which is deleterious for the environment. Therefore, they are eco-friendly.

Cheap and availability: DNA computers are inexpensive to build. All the materials are available because these materials are easily collected from nature.

Besides the above advantages, it has many other benefits like the presence of a distinct memory block that encodes DNA sequences. In addition, DNA computers will be lightweight if it is built up properly.

1.3 Demerits of DNA Computing

DNA computing has the following disadvantages:

It generates solution sets for some relatively simple problems. The high count of DNA strands generated means that an impractically large amount of memory may be required. Alternatively, a simple problem might require a lengthy DNA strand for encoding. Therefore, it becomes more challenging to solve problems of significant size because of the need for large amounts of DNA.

Another disadvantage is that errors are possible because of pair mismatches. That is why DNA computing has higher error rates.

The individual operations of DNA computers are relatively slower than silicon-based computers because of their capability for massive calculations.

The DNA computers may be destroyed gradually over time. For example, over a 6-month computation period, there is a chance that the DNA will degrade in the aqueous reaction mixture. As a result, DNA computers are prone to error.

DNA computers cannot replace traditional computers as they are not easily programmable. The average user cannot sit down at a familiar keyboard and get to work, limiting the reach of the technology.

DNA cryptography does not contain any mature mathematical background, whereas other cryptography systems have.

Another challenge is that response times of any operation can be slow — on the time-scale of minutes, hours, or even days. Thus DNA computers often take longer to find the correct solution to problems tractable by conventional computers.

1.4 Challenges of DNA Computing

DNA computing has several benefits such as parallel processing with very high efficiency, high storage capacity, and lower power consumption, but it also has several challenges. The practical operation of a DNA computer involves experimental errors that can increase with the number of steps. As a result, it needs human assistance or human intervention during the process. The correct answer can be missing while solving the more prominent size formulation.

Another problem is the need to represent real-world problems as DNA strands. Each datum has to be encoded in unique DNA strands because the DNA design of each datum is specific and cannot be used for other problems.

DNA design needs to consider other interactions such as the level of GC-content within the nucleotide sequence which might lead to unwanted secondary structures or incorrect bonding. This contradicts the design requirement of error-free biochemical operation. Such constraints reduce the design flexibility available for solving a problem.

In conclusion, DNA computing has many advantages over silicon-based computers, most notably the ability to perform millions of calculations simultaneously using molecules. Despite this, DNA computers have many limitations to handling real-world problems due to implementation issues.

1.5 Motivation for DNA Computing

The US engineer Gordan Moore observed that the number of transistors that can fit on a single silicon chip doubles roughly every two years. In other words, the cost of computers is halved every two years. As a result, as computers effectively shrink smaller and smaller, their performance has enjoyed exponential growth. Beyond Moore's Law, researchers started to think about different objectives and started the quest for novel ideas for processing the information. Finally, they found several exciting directions in the general area of unconventional computing, including research in quantum computing and molecular computing.

Generally, molecular computing is a branch of computing that uses DNA and the hardware of molecular biology. Instead of using silicon-based molecular computing, it is possible to pack vastly more molecular circuitry onto a microchip than traditional silicon. Their size measures on the nanometer scale, making it possible to manufacture chips containing billions and trillions of switches and components. Synthetic molecular systems have implemented Boolean logic operations, memory units, and arithmetic functions. Although these can perform basic Boolean operations and simple computations, these systems have limited complexity which is very challenging for building more sophisticated devices.

Moreover, these systems are still distant from the natural information processing within cells. If it is possible reproduce these natural processes for ourselves, building a biocomputer would become a reality. These would be particularly useful for some special applications such as solving complex combinatorial problems over traditional silicon-based computers due to the ability to run parallel computations. Future novel areas include sensing and innovative switchable materials controlled by bioelectronics devices, process signals controlled by external signals and signal controlled-release processes.

1.6 Summary

This chapter mainly covers the introduction of DNA computing, its pros and cons, its challenges and the motivation for DNA computing. It is now a most exciting area for researchers to explore. There are many opportunities for expanding on and manipulating the characteristics of DNA. DNA computers are capable of solving real applications, mainly industrial engineering and management engineering problems. If implemented properly, DNA computers will become an alternative that can solve the difficulties faced by current silicon-based computers.

Although DNA computing has become a more popular topic to try and understand, it is still a theoretical field. There are still some obstacles that are yet to be solved for building up a DNA computer. It is difficult to predict what directions the researchers will follow and what applications will be more efficient for DNA computing. Its computing model always depends on one molecular technique to solve various problems. Therefore at this point in time, it is tough to implement and has yet to provide a feasible alternative to cryptography security.

Chapter 2

Fundamental Operations in DNA Computing

2.1 Introduction

DNA computing is a field that uses DNA, biochemistry, and molecular biology rather than traditional silicon chips for computations. Recently, researchers have emphasized exploring this field because of several advantages of DNA computing. For example, Boolean logic uses two input or output states, and binary logic faces challenges from non-Boolean logic when processing uncertain or imprecise information. DNA computing is an emerging branch of computing that replaces traditional electronic computing with deoxyribonucleic acid (DNA), biochemistry, and molecular biology hardware. It may seem strange that computation may be done in a test tube using biological molecules rather than semiconductor chips.

Millions of natural supercomputers exist inside living organisms. DNA (deoxyribonucleic acid) molecules, which make up human genes, have the ability to execute calculations hundreds of times quicker than even the most powerful human-built computers. DNA could one day be incorporated into a computer chip to produce a "biochip" that will allow computers to run even quicker. Complex mathematical problems have already been solved using DNA molecules. This chapter presents how operations can be performed in DNA computing and the detail of DNA logic operations.

2.2 DNA Computing Operations

In classical computing, data storage device stores data by converting them into binary digits. But in DNA computing, instead of binary digits, information or data will now be kept in the form of the nitrogen bases **A**, **T**, **G**, and **C**. These bases will make sequences to store data, and encoding and decoding are required to perform the operation with that DNA sequences so that the outcomes become meaningful.

The capacity to generate short DNA sequences artificially allows these sequences to be used as inputs for algorithms. DNA has properties that allow it to be used to simulate classical logic processes. Single-stranded DNA naturally migrates toward complementary sequences to form double-stranded complexes, whereas double-stranded DNA wants to be in double-stranded form.

A program on a DNA computer is executed as a series of synthesizing, extracting, modifying, and cloning the DNA strands. Instead of using electrical impulses to represent bits of information, the DNA computer uses the chemical properties of DNA molecules by examining the patterns of combination or growth of the molecules or strings. DNA can do this through the manufacture of enzymes, which are biological catalysts that could be called the "software," used to execute the desired calculation. Enzymes do not function sequentially, working on one DNA at a time. Rather, numerous copies of the enzyme can act massively parallel on many DNA molecules concurrently. DNA computers work by encoding the problem to be solved in DNA's language: the base-four number system, which includes the base-four values **A**, **T**, **C**, and **G**, which is more than enough when compared to an electronic computer, which only requires two numbers, 0 and 1.

DNA has cut, copying, pasting, repairing, and many other operations, just like a CPU has addition, DNA sequence-shifting, logical operators, and so on, that allow it to accomplish even the most complex computations. The right sequences are sorted out using genetic engineering methods in a DNA computer, which computes in test tubes or on a glass slide coated in 24K gold.

2.3 Performing Fundamental Operations in DNA Computing

As mentioned earlier, the sequence of base patterns will store pieces of information. In two-valued (binary) DNA operations, consider the following stages:

ACCTAG = true, which is equivalent to binary "1"; and
TGGATC = false, which is equivalent to binary "0."

These base sequences represent data. With these base sequences, all fundamental circuits and operations will be done in the following sections.

2.3.1 *DNA NOT operation*

The basic binary logical operation is fully conversant to us. As a result, the logical function of binary logic operations does not need to be explained again. Only how to construct them in the DNA computer to perform DNA computing is discussed. Figure 2.1 shows the operational diagram of the DNA NOT operation. To design this, a test tube is needed with the DNA mixture, and the annealing temperature is less than 60 °C. To perform the operation, DNAse enzyme is needed. And the base sequence ACCTAG will be used.

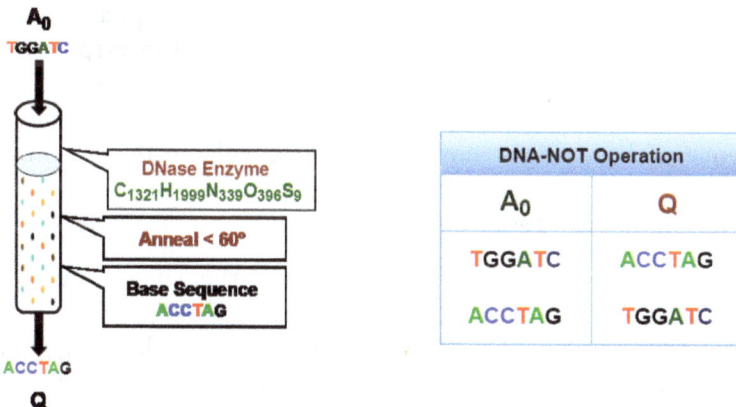

DNA-NOT Operation	
A_0	Q
TGGATC	ACCTAG
ACCTAG	TGGATC

Figure 2.1. Circuit architecture of DNA NOT operation.

Table 2.1.   Truth table of DNA
NOT operation.

A0	Q
TGGATC	**ACCTAG**
ACCTAG	**TGGATC**

A C C T A G
| | | | | |
T G G A T C

Figure 2.2.   The pair matching between the DNA base sequences.

The DNA NOT operation inverts the input base sequence. The operational table shows the input–output mapping in Table 2.1. Remember, the base sequences TGGATC and ACCTAG represent the Boolean false and true, respectively.

The operational logic is pretty straightforward. The DNA bases make pairs only if the sequences meet the conditions A-T or C-G. Here, the sequence ACCTAG is treated as the base sequence. Now if the input sequence makes a pair with the base sequence in the test tube as shown in Figure 2.2, then they will return the output as true. And if they do not make a pair, then the output will be false.

So, how to detect whether the input sequence and the base sequence make pairs or not? The answer will be detected by the DNAse enzyme.

The functional operations are as follows:

1. If the input base sequence is ACCTAG, then it will mix to the test tube where the other base sequence is also ACCTAG. Therefore, DNAse enzyme detects no base pair matching. Therefore, the result will be false, which represents the sequence TGGATC, as shown in Table 2.1.

2. If the input base sequence is TGGATC, then it will mix to the test tube where the other base sequence is ACCTAG. Therefore, DNAse enzyme will detect a base pair matching between them.

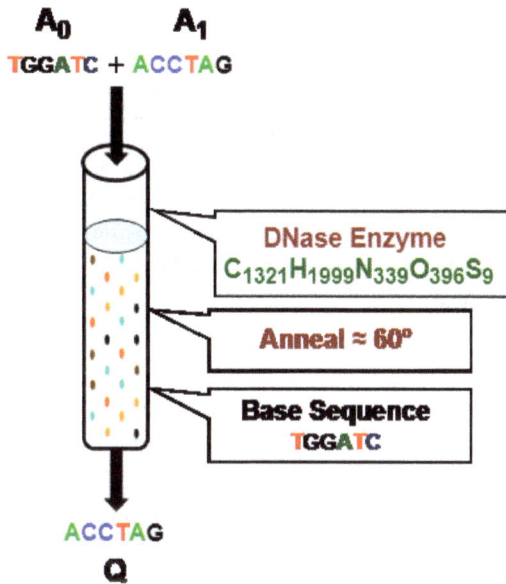

Figure 2.3. Circuit architecture of DNA OR operation.

Table 2.2. Truth table of DNA OR operation.

A1	A0	Q
TGGATC	TGGATC	TGGATC
TGGATC	ACCTAG	ACCTAG
ACCTAG	TGGATC	ACCTAG
ACCTAG	ACCTAG	ACCTAG

Therefore, the result will be true, which represents the sequence ACCTAG, as shown in Table 2.1.

This is how DNA NOT operation will invert the input.

2.3.2 *DNA OR operation*

Figure 2.3 shows the operational diagram of DNA OR operation. The logic is as same as the binary OR operation in classical computing. Input–output combinations of DNA OR operation logic are shown in Table 2.2.

To design the DNA OR operation, a test tube with the DNA mixture is needed, where the annealing temperature is 60 °C approximately. To perform the DNA OR operations, DNAse enzyme is needed. And here the base sequence TGGATC will be used.

Now the base sequence in the test tube is TCCATC. And the input to the DNA OR operation are two DNA sequences. Let the two input sequences be A_0 and A_1 and the output be Q.

Therefore, the working principles of DNA OR operation can be described as follows:

1. For A_0 = TGGATC and A_1 = TGGATC, the base sequence in the test tube is TGGATC. Therefore, no double-strand bond will be created. And thus, the DNAse enzyme will destroy them all. Therefore, the output will be TGGATC (which is equivalent to binary 0).

2. For A_0 = ACCTAG and A_1 = TGGATC, the base sequence in the test tube is TGGATC. Therefore, a double-strand bond will be created between ACCTAG and TGGATC. And one base sequence TGGATC will remain in the mixture which will be destroyed by the DNAse enzyme. Therefore, the output will be ACCTAG (which is equivalent to binary 1) because a bond has been created and DNAse enzyme will have no effect on them. So, the output result Q = ACCTAG for the given input.

3. For A_0 = TGGATC and A_1 = ACCTAG, the base sequence in the test tube is TGGATC. Therefore, a double-strand bond will be created between ACCTAG and TGGATC. And one base sequence TGGATC will remain in the mixture which will be destroyed by the DNAse enzyme. Therefore, the output will be ACCTAG because a bond has been created and DNAse enzyme will have no effect on them. So, the output result Q = ACCTAG again for the given input.

4. Now, for A_0 = ACCTAG and A_1 = ACCTAG, the base sequence in the test tube is TGGATC. Therefore, a double-strand bond will be created between ACCTAG and TGGATC. And one base sequence ACCTAG will remain in the mixture which will be destroyed by the DNAse enzyme because that sequence didn't create any bond with any other base sequence. Therefore, the output will be ACCTAG because a bond has been created and DNAse enzyme will have no effect on them. So, the output result Q = ACCTAG again for the given input.

Thus, the designed DNA OR operation produces the expected output for the given set of inputs.

2.3.3 *DNA NOR operation*

The NOR operation is nothing but the inverted output of the OR operation, both the DNA OR and DNA NOT operational systems are designed already, therefore, it's easy to design the operational system for the DNA NOR operation, which is shown in Figure 2.4.

From the above circuit, it is easy to understand the architecture of the DNA NOR operation. To perform DNA NOR operation, first DNA OR operation is needed to perform. Then the output of the DNA NOR operation will be inverted by the DNA NOT operation.

The input–output mapping for the DNA NOR operation is shown in Table 2.3. Consider the input–output mapping in Table 2.3, where the inputs are TGGATC and ACCTAG, and they produce TGGATC as output which is nothing but the inverted output from the DNA OR operation.

As the working procedures of DNA OR and DNA NOT operations have already been explained, it is easy to understand the working procedure of DNA NOR operation.

The functioning mechanism for the input sequence patterns to perform DNA NOR operation is given in the following:

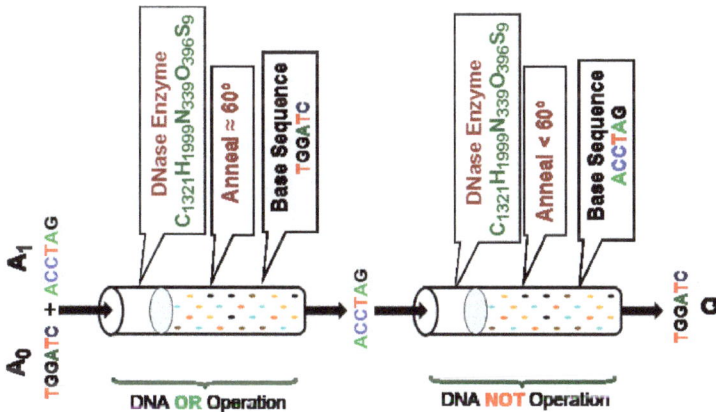

Figure 2.4. Circuit architecture of DNA NOR operation.

Table 2.3. Truth table of DNA NOR operation.

A1	A0	Q
TGGATC	TGGATC	ACCTAG
TGGATC	ACCTAG	TGGATC
ACCTAG	TGGATC	TGGATC
ACCTAG	ACCTAG	TGGATC

1. When the input base sequences both are TGGATC, then the DNA OR will generate TGGATC as output. And the DNA NOT operation will invert the output of the DNA OR operation. Therefore, the final output will be ACCTAG which is equivalent to binary 1.
2. When the input base sequences are TGGATC and ACCTAG, then the DNA OR will produce ACCTAG as output. And the DNA NOT will generate output TGGATC by inverting the output of the DNA OR operation. And the final output will be obtained as TGGATC.
3. When the input base sequences are ACCTAG and TGGATC, then the DNA OR will again produce ACCTAG as output. And the DNA NOT will generate output TGGATC by inverting the output of the DNA OR operation. And the final output will be obtained as TGGATC.
4. And when the input base sequences both are ACCTAG, then the DNA OR will again produce again ACCTAG as output. And the DNA NOT will generate output TGGATC by inverting the output of the DNA OR operation. And the final output will be obtained as TGGATC.

Therefore, the expected output for the given set of input base sequences has been obtained as shown in Table 2.3.

2.3.4 *DNA NAND operation*

In the structure of DNA OR, TGGATC is used as the base sequence in the test tube. The DNA computing system which will perform the operations of DNA NAND operation is containing ACCTAG as the base sequence instead of TGGATC. And the annealing temperature should be more than 60 °C to perform this operation.

Table 2.4. Truth table of DNA NAND operation.

A1	A0	Q
TGGATC	TGGATC	ACCTAG
TGGATC	ACCTAG	ACCTAG
ACCTAG	TGGATC	ACCTAG
ACCTAG	ACCTAG	TGGATC

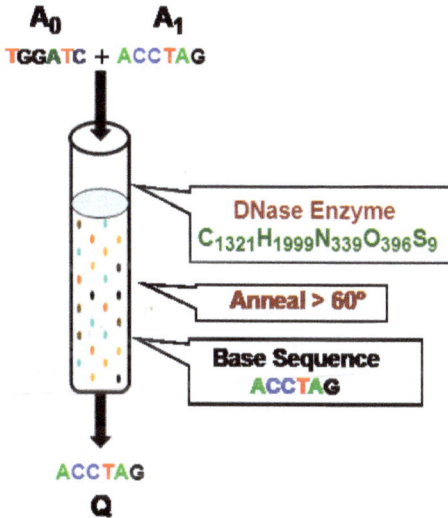

Figure 2.5. Circuit architecture of DNA NAND operation.

The output for the given input sequences that will be produced by the DNA NAND operation is shown in Table 2.4.

From Table 2.4, it is clear that the DNA NAND will produce an output sequence TGGATC only if the given input sequences both are ACCTAG. Otherwise, it will generate ACCTAG as an output always. Figure 2.5 shows the circuit architecture of the DNA NAND operation.

As always, let's take the set of input sequence patterns and observe the behavior of the system of the DNA NAND operation. Let the two input sequences be A_0 and A_1 and the output be Q.

Therefore, the working principles of DNA NAND operation can be described as follows:

1. For A_0 = TGGATC and A_1 = TGGATC, the base sequence in the test tube is ACCTAG. Therefore, a double-strand bond will be created between ACCTAG and TGGATC. And one base sequence TGGATC will remain in the mixture which will be destroyed by the DNAse enzyme. So, the output will be ACCTAG (which is equivalent to binary 1) because a bond has been created and DNAse enzyme will have no effect on them. Thus, the final output will be obtained as Q = ACCTAG for the given inputs.

2. For A_0 = ACCTAG and A_1 = TGGATC, the base sequence in the test tube is ACCTAG. Therefore, a double-strand bond will be created between ACCTAG and TGGATC. And one base sequence ACCTAG will remain in the mixture which will be destroyed by the DNAse enzyme. Therefore, the output will be ACCTAG because a bond has been created and DNAse enzyme will have no effect on them. So, the final output will be obtained as Q = ACCTAG for the given input sequences.

3. For A_0 = TGGATC and A_1 = ACCTAG, the base sequence in the test tube is ACCTAG. Therefore, a double-strand bond will be created between ACCTAG and TGGATC. And one base sequence ACCTAG will remain in the mixture which will be destroyed by the DNAse enzyme. So, the output will be again ACCTAG because a bond has been created and DNAse enzyme will not affect them. Thus, the final output will be obtained as Q = ACCTAG for the given input sequences.

4. Now, A_0 = ACCTAG and A_1 = ACCTAG, the base sequence in the test tube is ACCTAG. Therefore, no double-strand bond will be created among the base sequences. Consequently, all the base sequences will be destroyed by the DNAse enzyme because they didn't create any bond with any of the base sequences. So, the output will be TGGATC because no bond has been created (which means logically false). Thus, the final output will be obtained as Q = TGGATC for these given input sequences.

2.3.5   *DNA AND operation*

Normally, the AND operation is implemented first, then invert the output of the AND operation, and get the result of the NAND operation. But, in DNA computing, it is totally opposite.

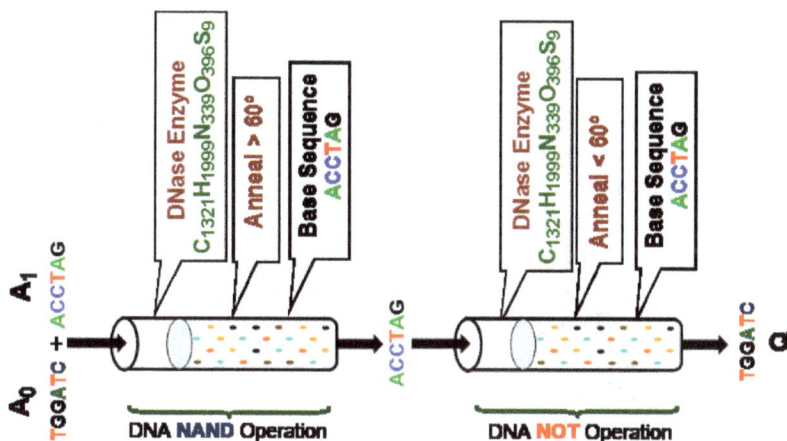

Figure 2.6. Circuit architecture of DNA AND operation.

Table 2.5. Truth table of DNA AND operation.

A1	A0	Q
TGGATC	TGGATC	TGGATC
TGGATC	ACCTAG	TGGATC
ACCTAG	TGGATC	TGGATC
ACCTAG	ACCTAG	ACCTAG

This is because it is easy to implement the NAND operation first in DNA computing. Then by inverting the output of NAND operations, it is easy to get the output of the AND operations. Figure 2.6 is to understand how to get the output of the DNA AND operations from the DNA NAND operations. Table 2.5 shows the truth table in DNA AND operation.

From Table 2.5, it is clear that the DNA AND operation will generate the output sequence ACCTAG only if the input base sequences both are ACCTAG; otherwise, the output sequence will always be TGGATC.

DNA AND operation is nothing but the inverted value of the DNA NAND operation. So first DNA NAND operation will perform and then DNA NOT operation will perform to invert the output value of the DNA NAND operation.

The functioning mechanism for the input sequence patterns to perform the DNA AND operation is given in the following:

1. When the input base sequences both are TGGATC, then the DNA NAND will generate ACCTAG as output, as shown in Table 2.4. And the DNA NOT operation will invert the output of the DNA NAND operation. Therefore, the final output will be obtained as TGGATC which is equivalent to binary 1.
2. When the input base sequences are TGGATC and ACCTAG, then the DNA NAND will produce ACCTAG as output. And the DNA NOT will generate output TGGATC by inverting the output of the DNA NAND operation. Thus, the final output will be obtained as TGGATC.
3. When the input base sequences are ACCTAG and TGGATC, then the DNA NAND will produce ACCTAG as output. And the DNA NOT will generate output TGGATC by inverting the output of the DNA NAND operation. So, the final output will be obtained as TGGATC.
4. When the input base sequences both are ACCTAG, then the DNA NAND will produce TGGATC as output. And the DNA NOT will generate output ACCTAG by inverting the output of the DNA NAND operation. Therefore, the final output will be obtained as ACCTAG.

2.3.6 *DNA XOR operation*

Table 2.6 shows the truth table of the DNA XOR operation. It will produce ACCTAG when the given input sequences are not the same sequence pattern. And when both the inputs are the same, then DNA XOR will generate TGGATC as output. To design the architecture

Table 2.6.   Truth table of DNA XOR operation.

A1	A0	Q
TGGATC	TGGATC	TGGATC
TGGATC	ACCTAG	ACCTAG
ACCTAG	TGGATC	ACCTAG
ACCTAG	ACCTAG	TGGATC

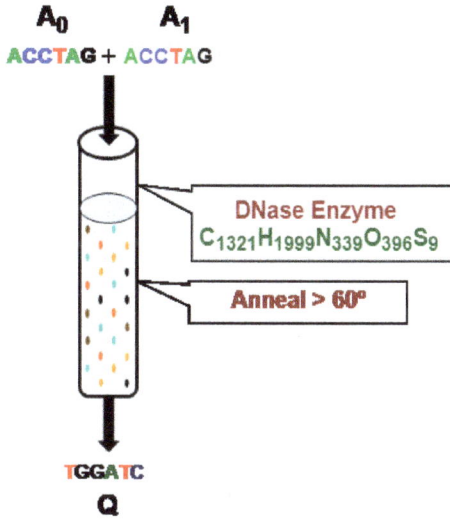

Figure 2.7. Circuit architecture of DNA XOR operation.

of the DNA XOR operation, there is no need for any base mixture as none of the sequences produces the DNA XOR output which is shown in Table 2.6. The input sequences must be complementary in order to have opposite values, and they will bind together to form a double-stranded sequence.

The logic is very simple. If the input sequences are the same, then they will not be able to make bonds together. If bonds are not created, the DNAse enzyme will destroy them, where the output will be false (TGGATC). When the input sequences are not the same, they will create the DNA double strands, and therefore DNAse enzyme will not affect them. As a result, the output will be true (ACCTAG). The architecture of the DNA XOR operation is shown in Figure 2.7.

In this case, there is no need for a base sequence to perform DNA XOR operations, where the required ideal annealing temperature is more than 60 °C.

2.3.6.1 *Working principles of DNA XOR operation*

The output provided by the designed DNA XOR system for the given set of input sequences is discussed in the following. Here the two inputs are A_0 and A_1 and the output is Q.

1. When input sequences A_0 and A_1 both are TGGATC, no DNA double-strand will form. So, the DNase enzyme will destroy the nitrogen bases in the mixture. Therefore, the output sequence will be TGGATC.
2. When input sequences A_0 is ACCTAG and A_1 is TGGATC, DNA double strands will form. So, the DNase enzyme will not affect the nitrogen bases in the mixture. Therefore, the output sequence will be ACCTAG.
3. When input sequences A_0 is TGGATC and A_1 is ACCTAG, DNA double strands will form. So, the DNase enzyme will not affect the nitrogen bases in the mixture. Therefore, the output sequence will be ACCTAG.
4. When input sequences A_0 and A_1 both are ACCTAG, then no DNA double strands will form. So, the DNase enzyme will destroy the nitrogen bases in the mixture. Therefore, the output sequence will be TGGATC.

So, the designed DNA XOR system provides the expected output results which are shown in the DNA XOR operation (Table 2.6).

2.3.7 *DNA XNOR operation*

The DNA XNOR is the inverted output of the DNA XOR operation. Therefore, a DNA NOT operational system should be added to the output of the DNA XOR operational system to get the output result of the DNA XNOR operation. Table 2.7 shows the operations in the DNA XNOR. Figure 2.8 shows the circuit architecture of the DNA XNOR operations.

Table 2.7.   Truth table of DNA XNOR operation.

A1	A0	Q
TGGATC	TGGATC	ACCTAG
TGGATC	ACCTAG	TGGATC
ACCTAG	TGGATC	TGGATC
ACCTAG	ACCTAG	ACCTAG

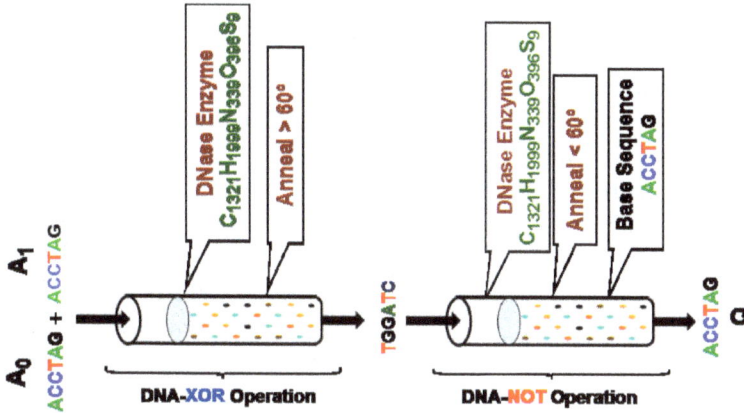

Figure 2.8. Circuit architecture of DNA XNOR operation.

The output provided by the designed DNA XNOR system for the given set of input sequences is given here. Assume that the two inputs are A_0 and A_1 and the output is Q.

1. When the input base sequences both are TGGATC, the DNA XOR will generate TGGATC as output which is shown in Table 2.6. And the DNA NOT operation will invert the output of the DNA XOR operation. Therefore, the final output will be ACCTAG as shown in Table 2.7 which is equivalent to binary 1.
2. When the input base sequences are TGGATC and ACCTAG, the DNA XOR will produce ACCTAG as output. And the DNA NOT will generate output TGGATC by inverting the output of the DNA XOR operation. So, final output will be TGGATC.
3. When the input base sequences are ACCTAG and TGGATC, the DNA XOR will produce ACCTAG as output. And the DNA NOT will generate output TGGATC by inverting the output of the DNA XOR operation. Thus, final output will be TGGATC.
4. And when the input base sequences both are ACCTAG, the DNA XOR will produce TGGATC as output. And the DNA NOT will generate output ACCTAG by inverting the output of the DNA XOR operation. Therefore, the final output will be ACCTAG.

So, the designed DNA XOR system provides the expected output results which are shown in the DNA XNOR operation Table 2.7.

2.4 Summary

The architecture of the DNA computer is not like the classical or quantum computer; it uses chemical reactions performed in a test tube to produce the output. The processes to perform DNA computation are preparing, mixing and annealing, melting, amplifying, separating, extracting, cutting, ligating, substituting, marking and destroying sequences, and detecting and reading sequences. In the truth table, the way DNA operations are performed is the same as classical computation. But the design of the circuit architecture is totally different. The total execution time is the required time to perform the largest pipeline in the operation. The heat required to perform a regular operation in DNA computing is fixed. The required heat is 284–490 °C. All basic logic operations with their architectures, truth tables, and working principles are presented in this chapter.

Bibliography

Kenneth J. Breslauer, Ronald Frank, Helmut Blöcker, and Luis A. Marky. Predicting DNA duplex stability from the base sequence. *Proceedings of the National Academy of Sciences*, 83(11): 3746–3750, 1986.

Susan M. Freier, Ryszard Kierzek, John A. Jaeger, Naoki Sugimoto, Marvin H. Caruthers, Thomas Neilson, and Douglas H. Turner. Improved free-energy parameters for predictions of RNA duplex stability. *Proceedings of the National Academy of Sciences*, 83(24): 9373–9377, 1986.

Junzo Watada. DNA computing and its application. In *Computational Intelligence: A Compendium*, pp. 1065–1089. Springer, 2008.

Takashi Yokomori, Satoshi Kobayashi, and Claudio Ferretti. On the power of circular splicing systems and DNA computability. In *Proceedings of 1997 IEEE International Conference on Evolutionary Computation (ICEC'97)*, pp. 219–224. IEEE, 1997.

Xuedong Zheng, Jing Yang, Changjun Zhou, Cheng Zhang, Qiang Zhang, and Xiaopeng Wei. Allosteric DNAzyme-based DNA logic circuit: Operations and dynamic analysis. *Nucleic Acids Research*, 47(3): 1097–1109, 2019.

Chapter 3

DNA Arithmetic Operations

3.1 Introduction

DNA computing, or computation with DNA molecules, has received a lot of attention in recent high-performance computing research as one of the non-silicon-based computing. The tremendous parallelism of DNA molecules allows us to solve combinatorial NP-complete problems in a polynomial number of steps, compared to exponential computation time on a silicon-based computer.

This chapter explains designing the arithmetic operations in DNA computing and discuss the design procedures and their working principles. The arithmetic operations are given as follows:

1. DNA half adder;
2. DNA full adder;
3. DNA molecular adder;
4. DNA BCD adder;
5. DNA carry skip adder;
6. DNA carry look ahead adder;
7. DNA half subtractor;
8. DNA full subtractor;
9. DNA multiplier;
10. DNA divider;
11. DNA comparator.

Arithmetic operations play a vital role in all kinds of computations in modern science. Let's start exploring the above operations in terms of DNA computing.

3.2 DNA Half Adder

A half adder is an adder that adds two DNA molecular sequences to produce a sum (S) and a carry (C). This adder can only add two molecular sequences at a time. Sequence **ACCTAG** represents the value "true" and Sequence **TGGATC** represents the value "false." The truth table of a half adder is given in Table 3.1.

From Table 3.1,

$$S = A' B + A B'$$

$$= A \text{ XOR } B; \text{ and}$$

$$Cout = A B.$$

3.2.1 *Block diagram of a DNA half adder circuit*

The DNA circuit for a half adder is created by several DNA operations. To design a half adder, one XOR operation and one AND operation are required. In this DNA circuit, one DNA XOR operation is needed to produce the SUM of the DNA half adder. Additionally, to generate the carry, one DNA AND operation is used. The block diagram of a DNA full adder circuit is illustrated in Figure 3.1.

Table 3.1. Truth table of DNA half adder.

Inputs		Outputs	
A	**B**	**S**	**Cout**
TGGATC	TGGATC	TGGATC	TGGATC
TGGATC	ACCTAG	ACCTAG	TGGATC
ACCTAG	TGGATC	ACCTAG	TGGATC
ACCTAG	ACCTAG	TGGATC	ACCTAG

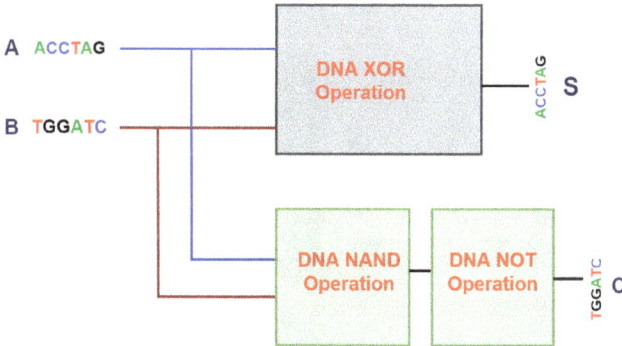

Figure 3.1.   Block diagram of DNA half adder.

3.2.2   *Circuit architecture*

Algorithm 3.1 presents the overall procedures of the DNA-based half adder operation. If the total number of inputs is n, then $O(n)$ is the run time complexity of this algorithm. DNA XOR operation is denoted by DO_DNA_XOR (**A, B**) and DNA AND operation is denoted by DO_DNA_AND (**A, B**). The "HALF_SUM" is used to represent the Half_Adder sum's output which is obtained after doing the DNA XOR operation. "HALF_Carry" considers the carry of the half adder which is obtained after doing DNA AND operation.

Algorithm 3.1 Half adder algorithm.

---

1. ***Begin***

2. ***while*** *i equals to 1 to n* ***do***

3. *HALF_SUM<-Do_DNA_XOR(Ai, Bi)*

4. *HALF_Carry<-Do_DNA_AND(Ai, Bi)*

5. ***end while***

Figure 3.2 depicts the DNA circuit of the half adder. In this half adder circuit, the XOR operation for A and B inputs is done by a DNA operation which generates an output S. For the output C, both inputs A and B pass through a DNA AND operation.

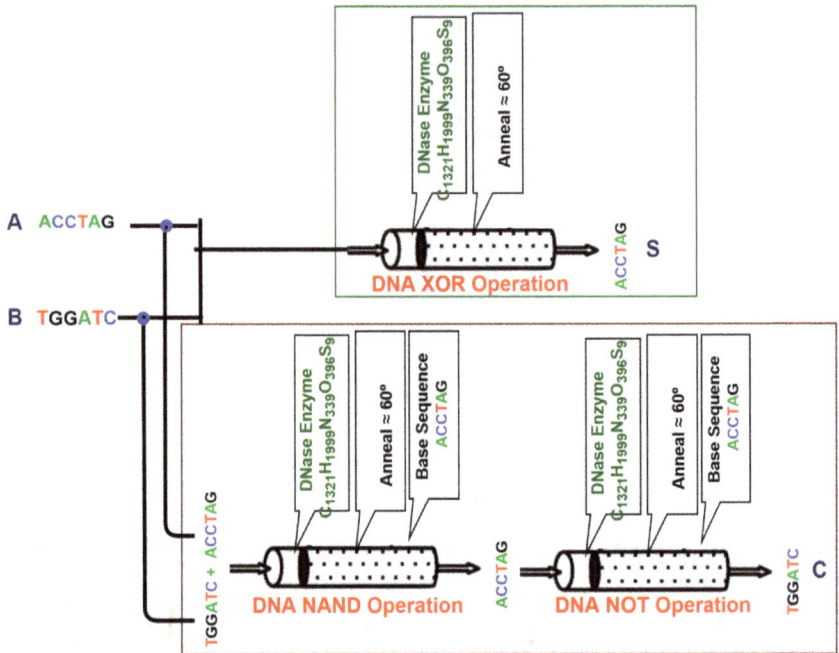

Figure 3.2.   DNA half adder circuit.

3.2.3   *Working principle*

The DNA half adder needs two inputs, A and B. For various values of inputs A and B, consider the following cases:

(i) When A and B are both "true" or **ACCTAG**, the output sequences of S is "false" or **TGGATC** and C is "true" or **ACCTAG**.

(ii) When A and B are both "false" or **TGGATC**, the output sequences of both S and C are "false" or **TGGATC**.

(iii) When A is "false" or **TGGATC** and B is "true" or **ACCTAG**, the output sequence (S) is "true" or **ACCTAG** and C is "false" or **TGGATC**.

(iv) When A is "true" or **ACCTAG** and B is "false" or **TGGATC**, the output sequence (S) is "false" or **TGGATC** and C is "true" or **ACCTAG**.

Table 3.2.   Truth table of a DNA full adder.

Inputs			Outputs	
A	**B**	**Cin**	**S**	**Cout**
TGGATC	TGGATC	TGGATC	TGGATC	TGGATC
TGGATC	TGGATC	ACCTAG	ACCTAG	TGGATC
TGGATC	ACCTAG	TGGATC	ACCTAG	TGGATC
TGGATC	ACCTAG	ACCTAG	TGGATC	ACCTAG
ACCTAG	TGGATC	TGGATC	ACCTAG	TGGATC
ACCTAG	TGGATC	ACCTAG	TGGATC	ACCTAG
ACCTAG	ACCTAG	TGGATC	TGGATC	ACCTAG
ACCTAG	ACCTAG	ACCTAG	ACCTAG	ACCTAG

3.3   DNA Full Adder

The DNA full adder is used to add three numbers including the carry input. The DNA full Adder solves the problem that occurs with DNA half adders. The DNA full adder is used to combine three 1-molecular sequence numbers A, B, and Cin. The full adder has three input states and two output states which are sum and carry. The molecular sequence ACCTAG represents the truth value "true" and the sequence **TGGATC** represents the false value "false." Table 3.2 shows the truth table of a DNA full adder.

From Table 3.2,

$$S = A' \ B'Cin + A' \ B \ Cin' + A \ B' \ Cin' + A \ B \ Cin$$
$$= Cin \ (A' \ B' + A \ B) + Cin' \ (A' \ B + A \ B')$$
$$= Cin \ XOR \ (A \ XOR \ B); \ and$$
$$Cout = A \ B + A \ Cin + B \ Cin \ (A + A')$$
$$= A \ B \ Cin + A \ B + A \ Cin + A' \ B \ Cin$$
$$= A \ B \ (1 + Cin) + A \ Cin + A' \ B \ Cin$$
$$= A \ B + A \ Cin + A' \ B \ Cin$$
$$= A \ B + A \ Cin(B + B') + A' \ B \ Cin$$

$$= A\ B\ Cin + A\ B + A\ B'\ Cin + A'\ B\ Cin$$
$$= A\ B\ (Cin + 1) + A\ B'\ Cin + A'\ B\ Cin$$
$$= A\ B + A\ B'\ Cin + A'\ B\ Cin$$
$$= A\ B + Cin\ (A'\ B + A\ B').$$

3.3.1   *Block diagram of a DNA full adder circuit*

The DNA circuit for a full adder is created by a number of DNA operations. To create a full adder, two XOR, two AND, and an OR operations are required. In this DNA circuit, two DNA XOR operations are required to produce SUM of the full adder. To produce the carry, two DNA AND operations and a DNA OR operation are needed. In a DNA circuit, a DNA NAND and a DNA NOT operations are required to achieve a DNA AND operation. Thus, instead of two DNA AND operations, two DNA NAND and two DNA NOT operations are used. The block diagram of a DNA full adder circuit is illustrated in Figure 3.3.

3.3.2   *Circuit architecture*

Algorithm 3.2 represents the overall procedures of the DNA-based full adder operation. To perform two DNA AND operations, two DNA XOR operations, and a DNA OR operation in the circuit, two

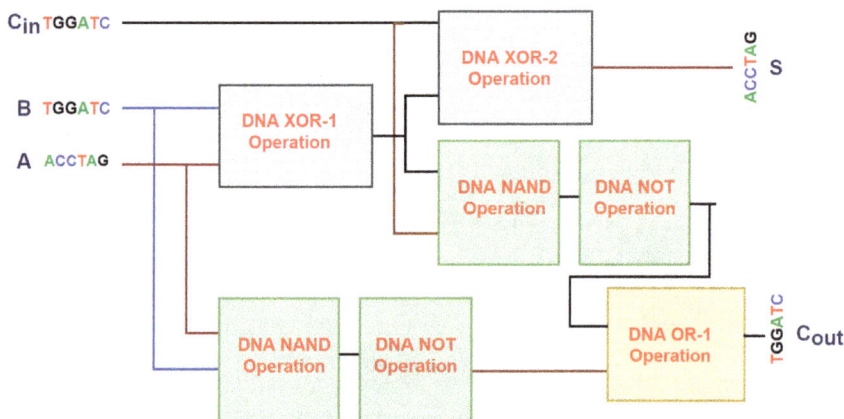

Figure 3.3.   Block diagram of a DNA full adder.

DNA XOR, two DNA AND, and one DNA OR operations are used. The DNA XOR operation is denoted by DO_DNA_XOR (X, Y), DNA AND operation is denoted by DO_DNA_AND (X, Y), and DNA OR operation is denoted by DO_DNA_OR (X, Y).

The "FA_SUM" is used to represent the Full_Adder sum's output which is obtained after doing the second DNA XOR operation. The "FA_Carry" is the final carry of the full adder which is obtained after doing DNA OR operation.

Algorithm 3.2 DNA full adder algorithm.

---

1. *Begin*

2. **while** *i equals to 1 to n* **do**

3. *Carry_1 <- DO_DNA_AND (A_i, B_i);*

4. *SUM_1 <- DO_DNA_XOR (A_i, B_i);*

5. *FA_SUM <- DO_DNA_XOR (SUM_1, C_i);*

6. *Carry_2 <- DO_DNA_AND (SUM_1, C_i);*

7. *FA_Carry <- DO_DNA_OR$(Carry\_1, Carry\_2)$;*

8. **end while**

Figure 3.4 shows the DNA circuit of the full adder. In this full adder, the first XOR operation for A and B input sequences is done by a DNA XOR operation and creates output A $\oplus$ B. This output, A $\oplus$ B, is used as one of the inputs of the second XOR operation and another input sequence comes from Cin. Finally, the output S of the full adder is generated through this DNA XOR operation which represents the sum or ACCTAG of the full adder. In addition, two DNA NAND operations and two DNA NOT operations are needed to get another output sequence of Cout. First DNA NAND operation occurs between inputs A and B and then the output passes through a DNA NOT operation to produce TGGATC. For the second DNA NAND operation, one input comes from DNA XOR operation, and another input is directly provided from Cin. The output of these two input sequences goes to a DNA NOT operation to generate TGGATC. Finally, the outputs of both DNA NOT operations pass through a DNA OR operation to produce the value of |Cout⟩ or TGGATC.

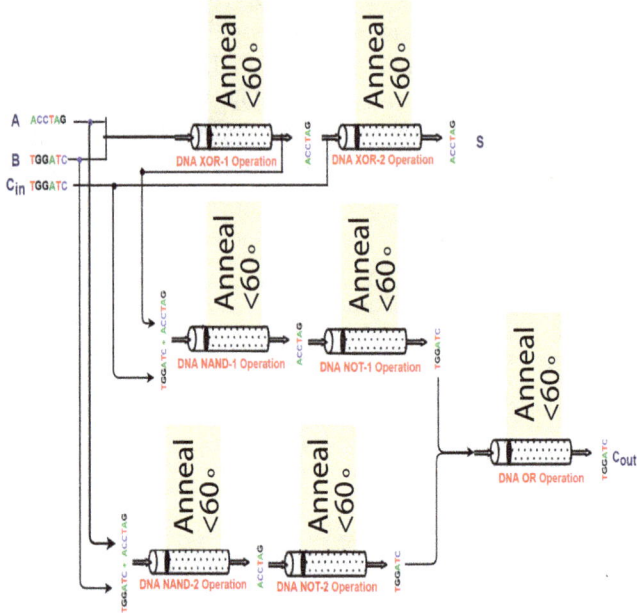

Figure 3.4.   DNA full adder circuit.

3.3.3   *Working principle*

The DNA full adder needs three inputs, A, B, and Cin, and provides two outputs S and Cout. To understand the working principle of a DNA full adder, consider the values of input A, B, and Cin which are **ACCTAG**, **TGGATC**, and **ACCTAG**, respectively. Then

(i) The output of DNA XOR-1 is "true" or **ACCTAG,** DNA AND-1 is "false" or **TGGATC** and DNA AND-2 is "false" or **TGGATC**.

(ii) Finally, the value of the output S is "true" or **ACCTAG,** and Cout is "false" or **TGGATC**.

3.4   **DNA Molecular Adder**

The DNA 2-molecular adder circuit requires three DNA XOR operations and three DNA AND operations which require three DNA NAND, and three DNA NOT operations, and one DNA OR

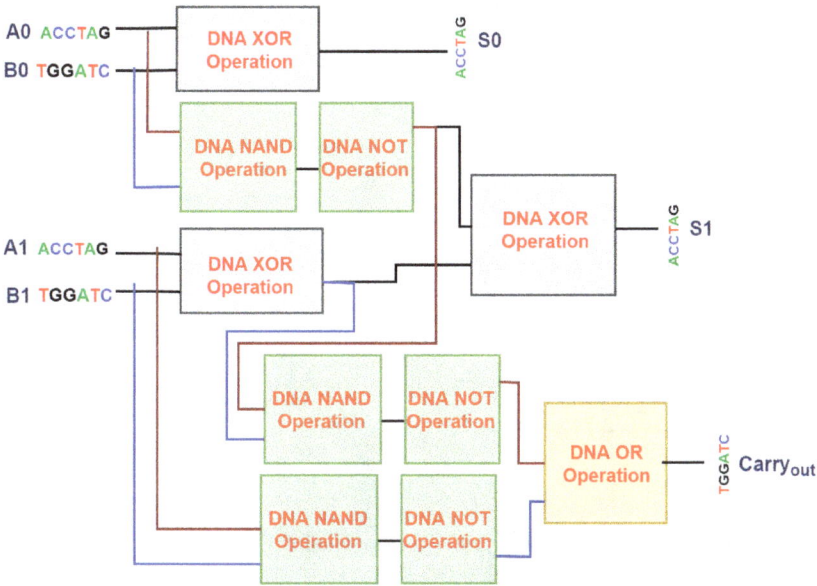

Figure 3.5. Basic block diagram of 2-molecular adder circuit.

operation in order to add the inputs of 2-molecular sequences A and B. Figure 3.5 illustrates the block diagram of the 2-molecular sequence adder circuit.

A DNA 2-molecular adder will add two 2-DNA sequence inputs and produce a 3-molecular sequence output. If A and B are the 2-molecular addend and augend molecular sequences where $A = A_1 A_0$ and $B = B_1 B_0$ respectively, then the addition operation of those inputs can be shown as

$$A = A_1 \, A_0$$
$$B = B_1 \, B_0$$
$$S = S_2 \, S_1 \, S_0,$$

where S is the result of the addition of the inputs and $S = S_2 \, S_1 \, S_0$ and the value can be either ACCTAG or TGGATC.

Figure 3.6 shows the general organizations of a DNA 2-molecular adder. Figure 3.6 shows that the block diagram of DNA 2-molecular adder has one DNA half adder operation and one DNA full adder

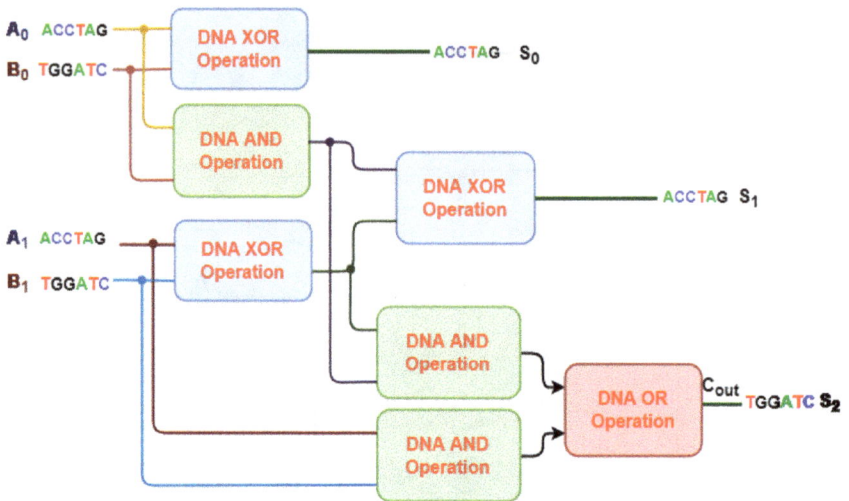

Figure 3.6.   Block diagram of a 2-molecular DNA adder circuit.

operation, where the carry output of the first DNA half-adder works as a carry input to the next DNA full adder.

Here, DNA AND operation is performed by a DNA NAND operation followed by a DNA NOT operation. Therefore, the block diagram of a DNA 2-molecular adder can be shown as in Figure 3.6.

According to the block diagram, it is easy to build the circuit architecture of a 2-molecular adder in DNA computing. The 2-molecular DNA adder circuit is given below in Figure 3.7.

3.4.1   *Working principle*

In the DNA, the 2-molecular DNA adder circuit shown in Figure 3.7, where A0 and B0 inputs perform a DNA XOR operation to generate the output of S0. These inputs also work as inputs of another DNA NAND operation which is a combination of DNA AND and DNA NOT operations. The output of this operation acts as an input for both DNA XOR-2 operation and DNA AND-2 operation. On the other hand, the inputs A1 and B1 pass through the DNA XOR-3 operation to produce ($A1 \oplus B1$). This ($A1 \oplus B1$) then goes to

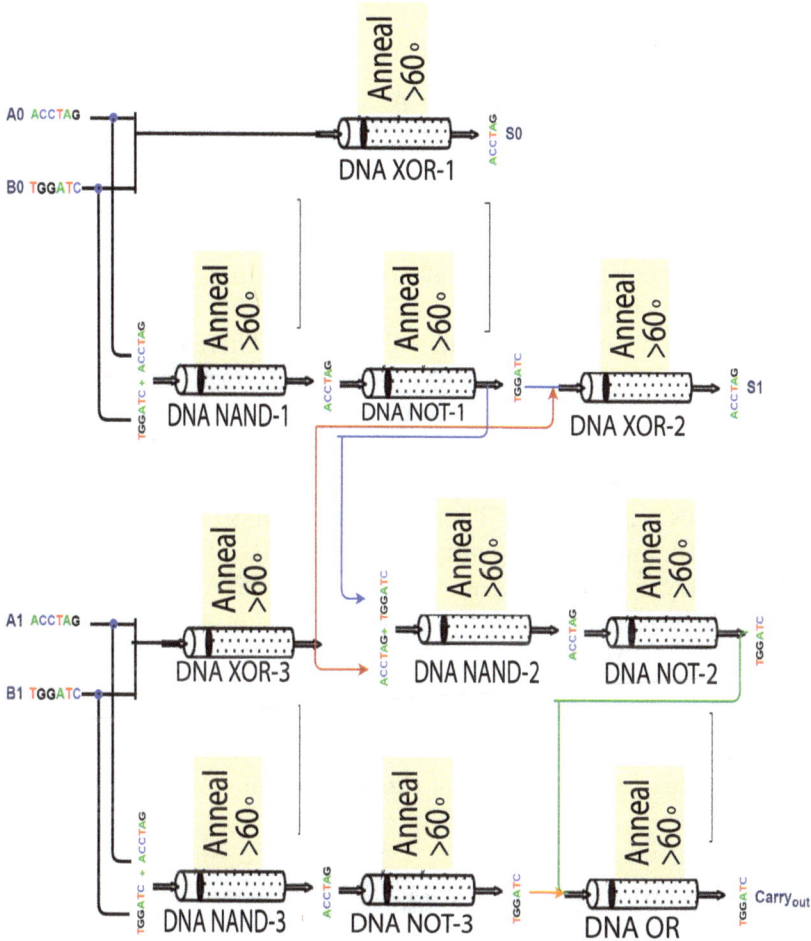

Figure 3.7.   2-Molecular DNA adder circuit.

the DNA XOR-2 operation and DNA AND-2 operation as the second input. Now, the second XOR then generates the output of S1. The output produced by the DNA NAND-2 operation goes to the DNA OR process whose other input comes from the DNA AND-3 operation of inputs A1 and B1. Finally, DNA OR operation gives the output of Cout which is the most significant molecular sequence of the addition result.

The circuit consists of a half adder followed by a DNA full adder and the operations for several inputs can be shown as follows:

1. Let, A1 = ACCTAG, A0 = ACCTAG, B1 = ACCTAG, and B0 = TGGATC. The values of A0 and B0 will work as the inputs of the DNA half-adder. Therefore, the sum S0 will be ACCTAG (binary 1), and the carry output will be TGGATC (binary 0). This carry will work as an input to the DNA full adder along with the inputs A1 and B1. Therefore, the full adder will produce TGGATC as the sum which is the value of S1, and ACCTAG as the carry output which is the value of S2. Therefore, S2 = ACCTAG, S1 = TGGATC, and S0 = ACCTAG which are the expected result of the addition operation for the given inputs.
2. Let, A1 = ACCTAG, A0 = TGGATC, B1 = ACCTAG, and B0 = TGGATC. The values of A0 and B0 will work as the inputs of the DNA half-adder. Therefore, the sum S0 will be TGGATC (binary 0), and the carry output will be TGGATC (binary 0) also. This carry will work as an input to the DNA full adder along with the inputs A1 and B1. Therefore, the full adder will produce TGGATC as the sum which is the value of S1, and ACCTAG as the carry output which is the value of S2.

Therefore, S2 = ACCTAG, S1 = TGGATC, and S0 = TGGATC which are the expected results of the addition operation for the given inputs.

3.4.2   4-Molecular DNA adder

For a DNA 4-molecular adder, the overall circuit can be divided into four parts as each sequence of input A performs a full adder operation (also a half adder operation can be performed for the first two input sequences) with each molecular sequence of B that has the same position. For 4-molecular addition, the inputs A and B consist of 4-molecular sequences such as

$$A = A_3 A_2 A_1 A_0 \quad \text{and} \quad B = B_3 B_2 B_1 B_0.$$

The general circuit of the DNA 4-molecular adder is shown in Figure 3.8.

From Figure 3.8, to perform a 4-molecular DNA addition, four DNA full adders are needed, where the carry output of each full adder

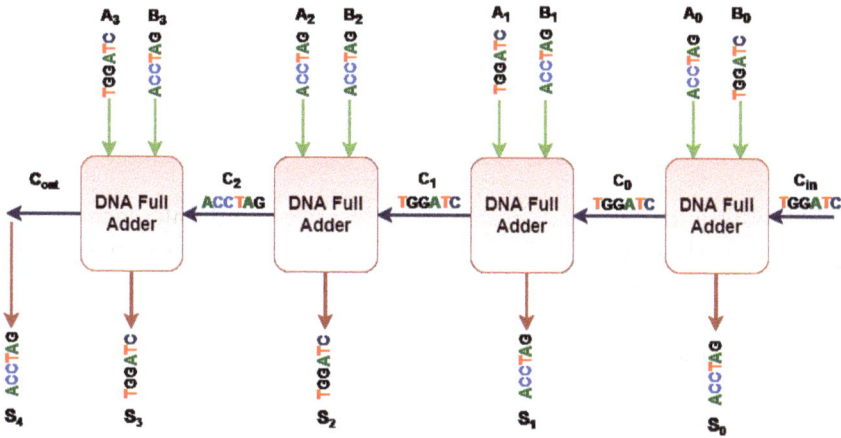

Figure 3.8. The 4-molecular DNA adder circuit.

will work as the carry input to the very next DNA full adder. And each full adder will generate the sum output of the addition of given inputs.

Eventually, four DNA full adder will generate S_0, S_1, S_2, and S_3. And the most significant molecular sequence of the result will be the carry output of the last DNA full adder which is S4.

Figure 3.8 shows the input–output illustrations, where inputs A and B are as follows:

A0 = ACCTAG, A1 = TGGATC, A2 = ACCTAG,

and A3 = TGGATC; and

B0 = TGGATC, B1 = ACCTAG, B2 = ACCTAG,

and B3 = ACCTAG.

This suggests that, A = 0101 and B = 1110.

After performing addition operation using the DNA 4-molecular adder, the output is as follows:

S_4 = ACCTAG, S_3 = TGGATC, S_2 = TGGATC, S_1 = ACCTAG, and S_0 = ACCTAG (i.e. S = 10011) which are the expected outputs.

Similarly, it is easy to construct an adder circuit for larger molecular DNA sequence inputs. For an n-molecular DNA adder, n DNA full adder will be needed which are connected in a manner where the carry outputs of each DNA full adder will work as an input to the next DNA full adder.

3.5 DNA BCD Adder

The BCD adder is used in computer systems and calculators that perform arithmetic operations in the decimal number system directly. The binary-coded form of decimal numbers is accepted by the BCD adder. A threshold of nine inputs and five outputs are required for the decimal adder.

3.5.1 *Block diagram of a DNA BCD adder*

For the construction of the BCD adder, two 4-molecular adders are required, along with two DNA AND operations (a combination of DNA NAND and DNA NOT) and two DNA OR operations. The block diagram of the BCD adder is given in Figure 3.9.

3.5.2 *Circuit architecture of a DNA BCD adder*

Figure 3.10 shows the DNA circuit of the BCD Adder. Here, the four outputs of the first 4-molecular DNA adder directly transfer to

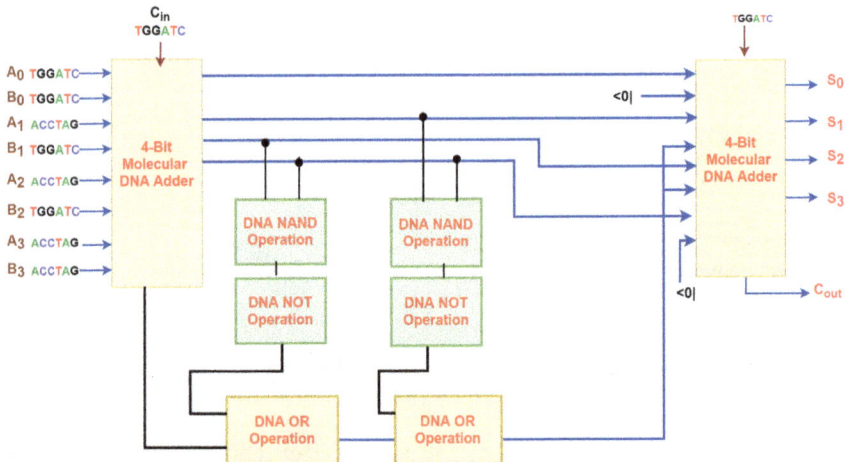

Figure 3.9.   Block diagram of DNA BCD adder circuit.

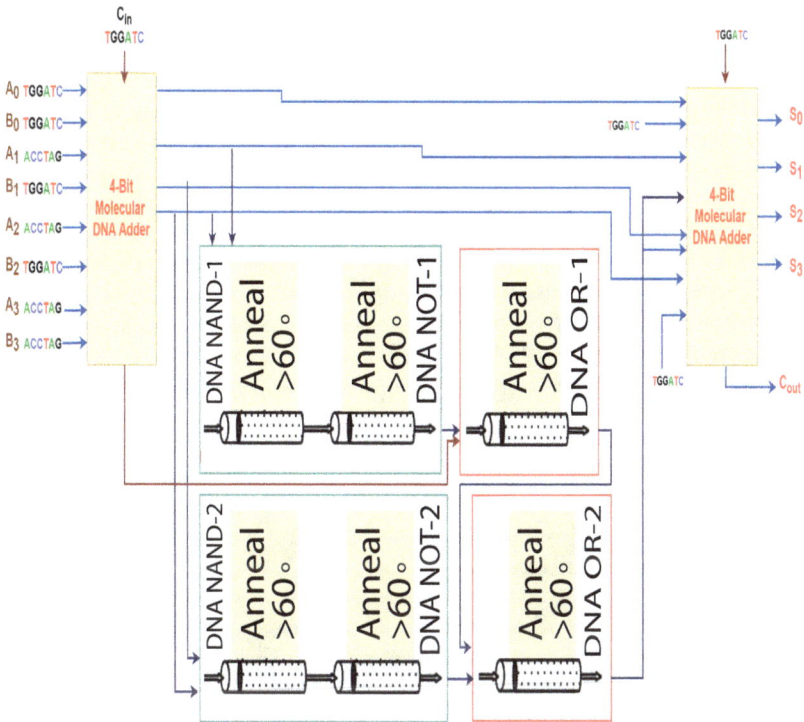

Figure 3.10. DNA BCD adder circuit.

the next 4-molecular DNA adder, as the inputs of A. However, in-between, two DNA AND operations and two DNA OR operations are performed their activities. The values of S1 and S3 perform the AND-1 operation, whereas, the DNA AND-2 operation is performed on S2 and S3. In the next step, both the outputs of the AND operation along with the Cout of the first 4-molecular DNA adder, sequentially execute two OR operations. Then, the result of the DNA OR-2 operation acts as the inputs for the B of the last 4-molecular DNA adder.

3.5.3 *Working principle*

The DNA BCD Adder needs two 4-molecular sequence adders where there are two 4-molecular sequence inputs A and B and a carry input, Cin. To understand the working principle of the DNA BCD Adder, consider the values of inputs A0, A1, A2, and A3 which are **ACC-TAG, TGGATC, TGGATC**, and **ACCTAG**, respectively equal to the decimal value "6" and B0, B1, B2, and B3 are **TGGATC, ACCTAG, TGGATC**, and **ACCTAG**, respectively equal to decimal value "5." Then

(i) The outputs of the first 4-molecular sequence adder will be S0 = **TGGATC**, S2 = **TGGATC**, S2 = **ACCTAG**, S3 = **TGGATC**, and |Cout⟩ = **ACCTAG**.

(ii) Now, the results of two DNA AND operations are S2. S3 = **ACCTAG and** S1. S3 = **TGGATC**.

(iii) By continuing the above result, the last DNA OR operation generates "True Value" or **TGGATC**.

(iv) Finally, the outputs of the second 4-molecular sequence adder produce are S0 = **TGGATC**, S2 = **ACCTAG**, S2 = **ACCTAG**, S3 = **ACCTAG**, and Cout = **TGGATC** which is 11 in the BCD number.

3.6 DNA Carry-Lookahead Adder

To reduce propagation delay, the DNA carry-lookahead adder is used. A brief discussion has already been done about this in the earlier chapter. Now, this section describes the construction of the DNA carry-lookahead adder. In the DNA carry-lookahead adder, the calculation of all the carry DNA sequences is done before the execution of the adder operations. The first carry DNA sequence is used for all other carry DNA sequences. As a result, no adder has to wait for the carry from the previous adder. This is the main technique which is used to construct a DNA carry-lookahead adder.

3.6.1 *Block diagram of a DNA carry-lookahead adder*

To calculate the carry part, a combinational circuit is needed that is constructed with DNA OR and DNA AND operations. So, the combinational circuit needs to be constructed in the DNA system. Then the DNA combinational circuits along with the DNA parallel adders will work as the DNA carry-lookahead adder. Figure 3.11 shows the general architecture of a 4-molecular DNA carry-lookahead

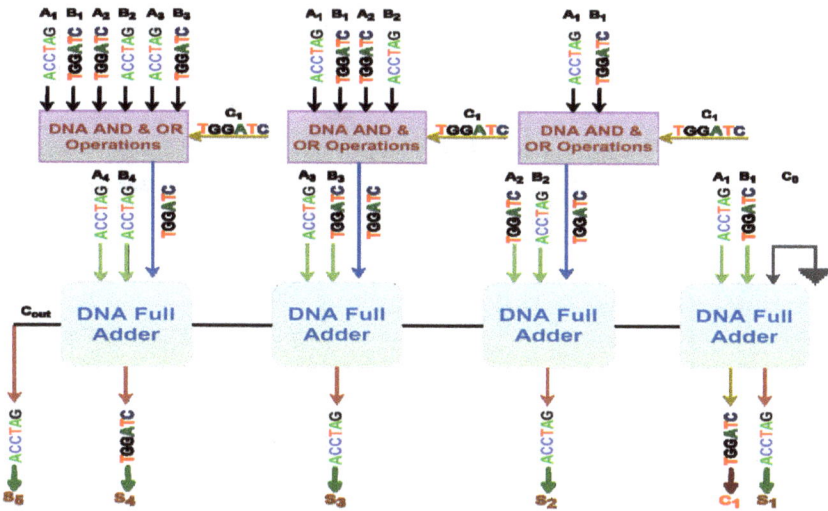

Figure 3.11. Block diagram of 4-molecular DNA carry-lookahead adder.

adder and Figure 3.12 shows the block diagram of a n-molecular DNA carry-lookahead adder.

The values of the carry part can be determined as follows:

$$C_2 = P_1 C_1 + G_1$$
$$C_3 = P_2 P_1 C_1 + P_2 G_1 + G_2$$
$$C_4 = P_3 P_2 P_1 C_1 + P_3 P_2 G_1 + P_3 G_2 + G_3$$
$$\cdots$$
$$C_n = P_{n-1} P_{n-2} \cdots \cdot P_2 P_1 C_1 + P_{n-1} P_{n-2} \cdots \cdot$$
$$P_3 P_2 \, G_1 + P_{n-1} P_{n-2} \cdots \cdot P_4 P_3 \, G_2 + \cdots$$
$$+ P_{n-1} P_{n-2} G_{n-3} + P_{n-1} G_{n-2} + G_{n-1},$$

where

$$P_i = A_i \textbf{ DNA-XOR } B_i,$$
$$G_i = A_i \textbf{ DNA-AND } B_i,$$

and

The sum $S_i = P_i \textbf{ DNA-XOR } C_i,$
The carry $C_{i+1} = (P_i \textbf{ DNA-AND } C_i) \textbf{ DNA-OR } G_i.$

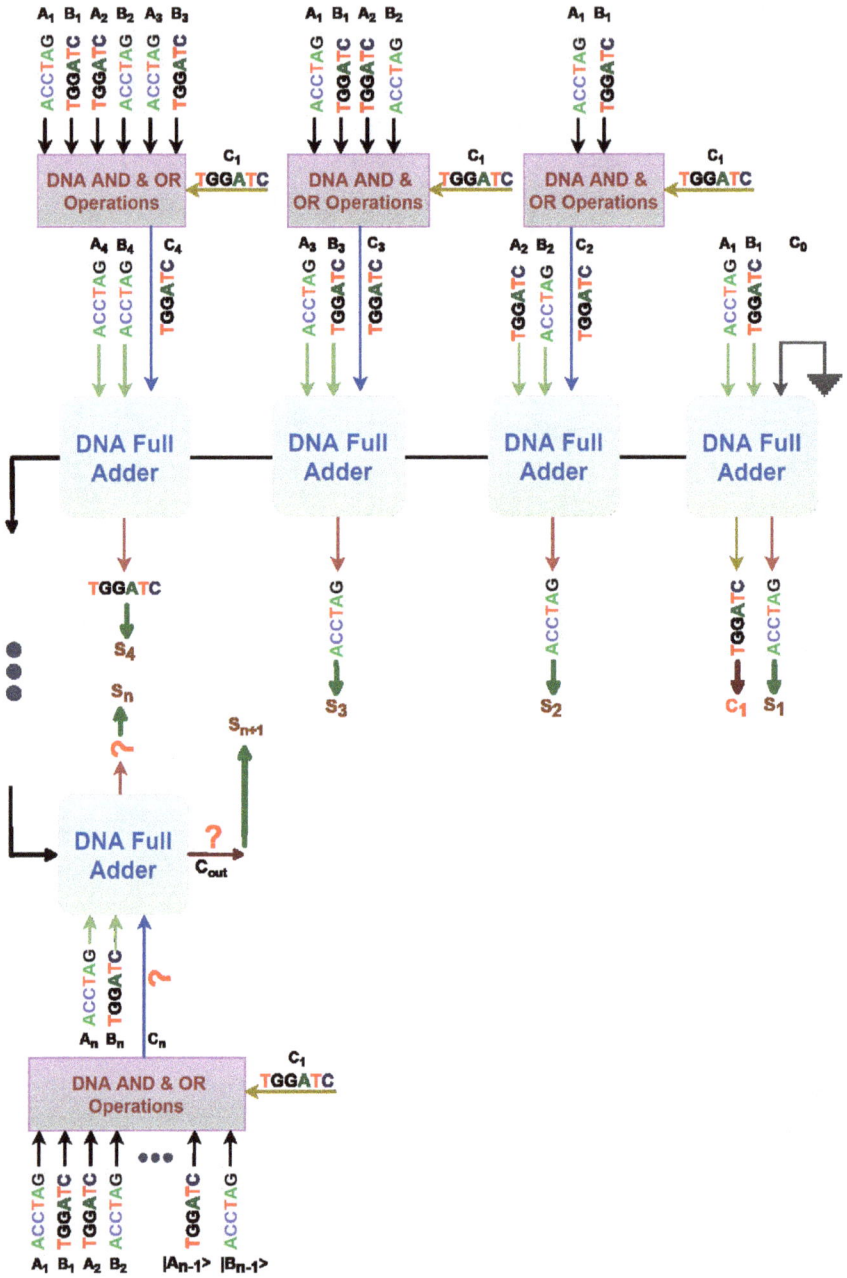

Figure 3.12.   Block diagram of *n*-molecular DNA carry-lookahead adder.

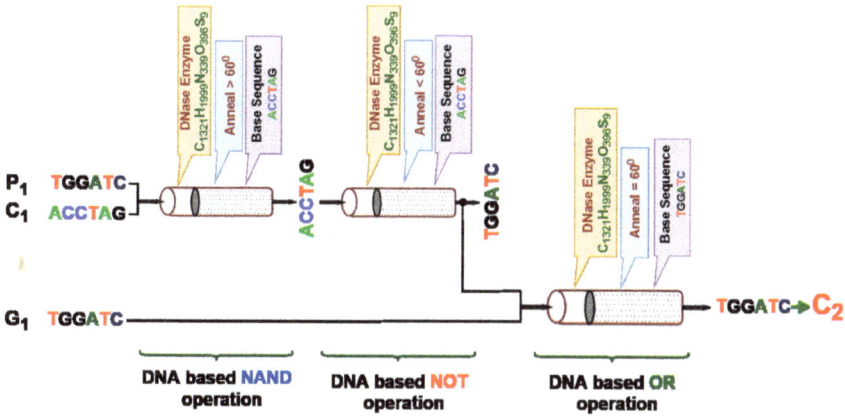

Figure 3.13.   Circuit architecture for deriving the value of C_2.

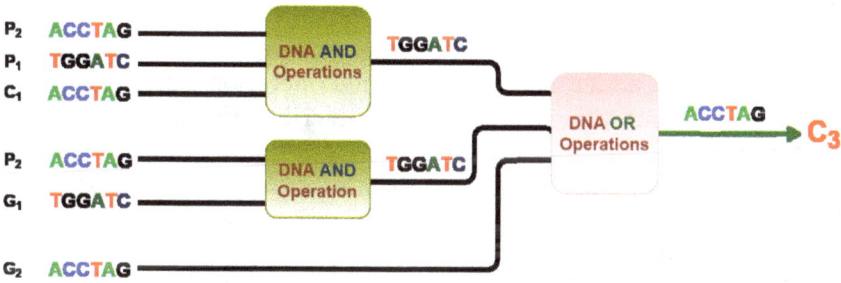

Figure 3.14.   General architecture for deriving the value of C_3.

Figure 3.13 shows the architecture of a DNA system to derive the value of C_2. If the value of $P_1 =$ TGGATC, $C_1 =$ ACCTAG, and $G_1 =$ TGGATC, then the value of C_2 will be TGGATC.

The general architecture to derive the value of C_3 is shown in Figure 3.14. It is seen that, if the value of $P_1 =$ TGGATC, $P_2 =$ ACCTAG, C1 = ACCTAG, $G_1 =$ TGGATC, and $G_2 =$ ACCTAG, then the value of C_3 will be ACCTAG.

It is important to note that the value of C_2 is not needed to determine the value of C_3. Accordingly, no need of the value of any carry output to determine the value of the carry from any DNA full adder.

3.6.2   *Circuit architecture*

In this section, a circuit diagram for the 3-molecular DNA carry-lookahead adder will be constructed as shown in Figure 3.15. First, a DNA full adder will be designed, and then the architecture of deriving P_1, and G_1, and now, using them, it is necessary to design the architecture to derive the value of C_2. Finally, a DNA full adder is constructed which will add the second DNA sequence of two inputs and get carry input from the value of C_2.

Now, the sum output is obtained, and in the same way, the architectures to derive P_2, G_2, and C_3 are designed. At this stage, another DNA full adder adds the most significant DNA sequences. And finally, the sum output and carry output are obtained. Note that in the architecture, C_1 is only used to get the value of C_2 and C_4.

3.6.3   *Working principles*

Consider an example to explain the working procedure of DNA carry-lookahead adder. Here, the addition operation is performed for 3-molecular input values. Figure 3.15 describes the working procedure as follows.

Suppose, the 3-molecular inputs are $A = A_3A_2A_1$ and $B = B_3B_2B_1$. Let, $A = 110$ and $B = 100$. After encoding, inputs A, $A_3 = ACCTAG$, $A_2 = ACCTAG$, and $A_1 = TGGATC$. And inputs B, $B_3 = ACCTAG$, $B_2 = TGGATC$, and $B_1 = TGGATC$.

Now, the addition operations are performed by using the DNA 3-molecular carry-lookahead adder and its working principle is discussed here.

1. At first, the addition is performed between A_0 and B_0, where $A_1 = TGGATC$ and $B_1 = TGGATC$. And the carry input sequence is also TGGATC for the first operation.

   (a) The first DNA full adder gets inputs TGGATC from all three inputs A_1, B_1, and C_{in}. Therefore, it produces TGGATC as the sum output. And it generates also TGGATC as the carry output. This is the value of C_1 which will be used to determine all the values of C_i.

   (b) The value of P_1 is TGGATC and the value of G_1 is also TGGATC. Therefore, $C_2 = TGGATC$ (P1C1 + G1).

   (c) So, $S_1 = TGGATC$ and $C_2 = TGGATC$.

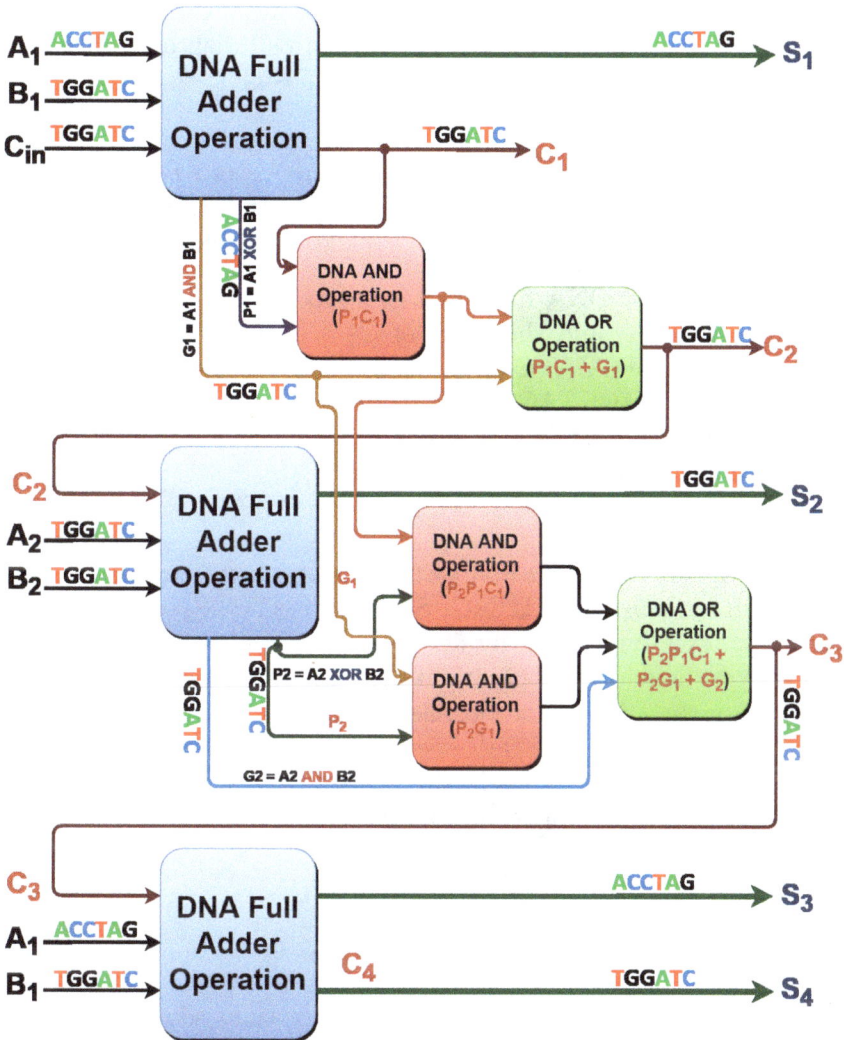

Figure 3.15.   3-Molecular DNA carry-look ahead adder.

2. Now, the addition is performed between A_2 and B_2. And here, A_2 = ACCTAG and B_2 = TGGATC. And the carry input sequence is also TGGATC (the value of C_2 from Step 1(b)).

   (a) The second DNA full adder gets inputs ACCTAG from A_2, TGGATC from both B_2, and C_2. Therefore, it produces ACCTAG as the sum output.

(b) The value of P_2 is ACCTAG and the value of G_2 is TGGATC. Therefore, $C_3 = $ TGGATC (see the circuit diagram, where the logical expression is $P_2P_1C_1 + P_2G_1 + G_2$).

(c) So, $S_2 = $ ACCTAG and $C_3 = $ TGGATC.

3. Now, the addition is performed between A_3 and B_3. And here, $A_3 = $ ACCTAG and $B_3 = $ ACCTAG. And the carry input sequence is again TGGATC (the value of C_3 from Step 2(b)).

(a) The second DNA full adder gets inputs ACCTAG from both A_3, and B_3 and the value of C_3 is TGGATC. Therefore, it produces TGGATC as the sum output. Thus, the final carry output is ACCTAG.

(b) This final carry is the most significant DNA sequence of the addition operation's result.

(c) Therefore, $S_3 = $ TGGATC and $S_4 = $ ACCTAG.

Now, after performing the operation in DNA carry-lookahead adder, the results are $S_4 = $ ACCTAG, $S_3 = $ TGGATC, $S_2 = $ ACCTAG, and $S_1 = $ TGGATC. And by decoding these values, it is necessary to perform the addition operations of two input values A = 110 and B = 100 in the DNA carry-lookahead adder which produce the sum S = 1010 and this sum is the expected result. And, it is also proved that all addition operations in the DNA carry-lookahead adder are performed using the carry outputs with only the value of the first generated carry output.

3.7 DNA Carry-Skip Adder

The DNA carry-skip adder skips several molecular sequences (often called block) if the addition process of the input molecular sequence does not generate any carry output which is equivalent to ACCTAG (binary equivalent "1"). It bypasses the first carry molecular sequences of the block when it is found that there exists no operation which generates carry output ACCTAG. And this is why it is also called the DNA bypass adder. Figure 3.16 shows how a carry is propagated through the operations.

Here, BP stands for block propagate which determines whether a carry will generate as ACCTAG or not. If there will be no carry

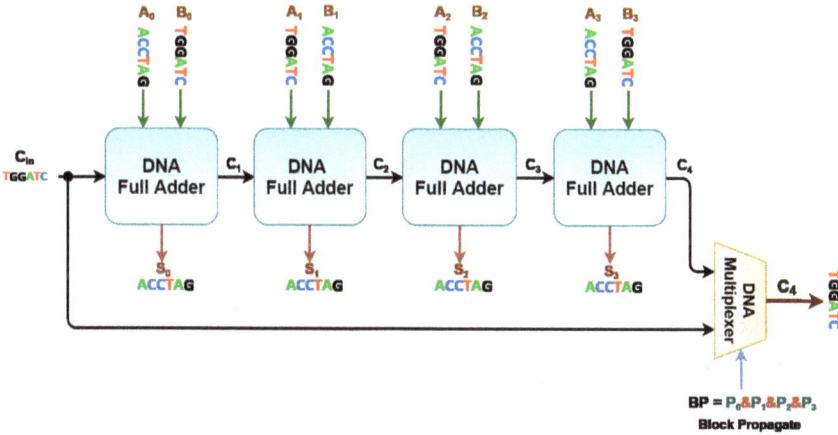

Figure 3.16. Basic structure of a 4-molecular DNA carry-skip adder.

output as ACCTAG, a DNA multiplexer will bypass the first carry input C_{in} as the input for the next DNA full adder operation. And, if there is the generated carry as ACCTAG within the four DNA full adder operations, this carry skip adder works just like a DNA molecular adder.

The block propagate BP can be determined as follows:

$$BP_i = P_0 \cdot P_1 \cdot P_2 \cdot \cdots \cdot P_i,$$

where $P_i = (A_i \textbf{ DNA-XOR } B_i)$.

3.7.1   *Block diagram*

Figure 3.17 shows the general architecture of a 4-molecular DNA carry-skip adder which is designed based on the working mechanism as shown in the previous section.

In Figure 3.17, the data is propagated if BP = ACCTAG. Then the circuit will bypass the first carry molecular sequence to the next DNA full adder using a DNA 2-to-1 multiplexer.

3.7.2   *Circuit architecture*

It is easy to construct the circuit architecture of a DNA carry-skip adder from the general architecture which is shown in the

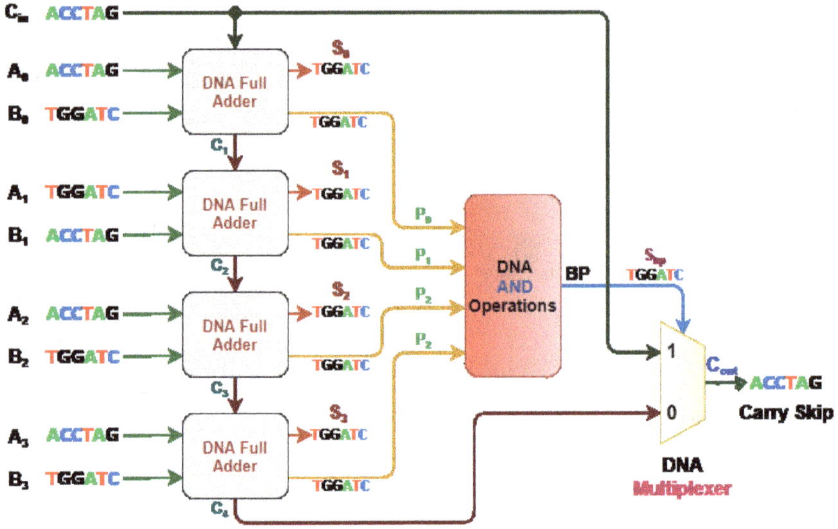

Figure 3.17.   Block diagram of a 4-molecular DNA carry-skip adder.

previous section. Figure 3.18 presents the complete architecture of a DNA 3-molecular carry-skip adder.

It is clear from Figure 3.18 that three DNA full adders and three DNA AND operations are required to determine the block propagate, and a 2-to-1 DNA multiplexer is used to bypass the block propagate which is necessary to build the 3-molecular DNA carry-skip adder.

3.7.3   *Working principles*

In the design of a 3-molecular DNA carry-skip adder, suppose $A = 101$ and $B = 010$. So, after encoding, A_0 = ACCTAG, A_1 = TGGATC, and A_2 = ACCTAG. And B_0 = TGGATC, B_1 = ACCTAG, and B_2 = TGGATC. It works as follows:

1. The first DNA full adder receives inputs A_0 = ACCTAG, B_0 = TGGATC, and C_0 = TGGATC. Therefore, the sum is ACCTAG, and the value of P_0 is ACCTAG.
2. For the second DNA full adder operation, input DNA sequences are A_1 = TGGATC and B_1 = ACCTAG. Therefore, P_1 = ACCTAG.

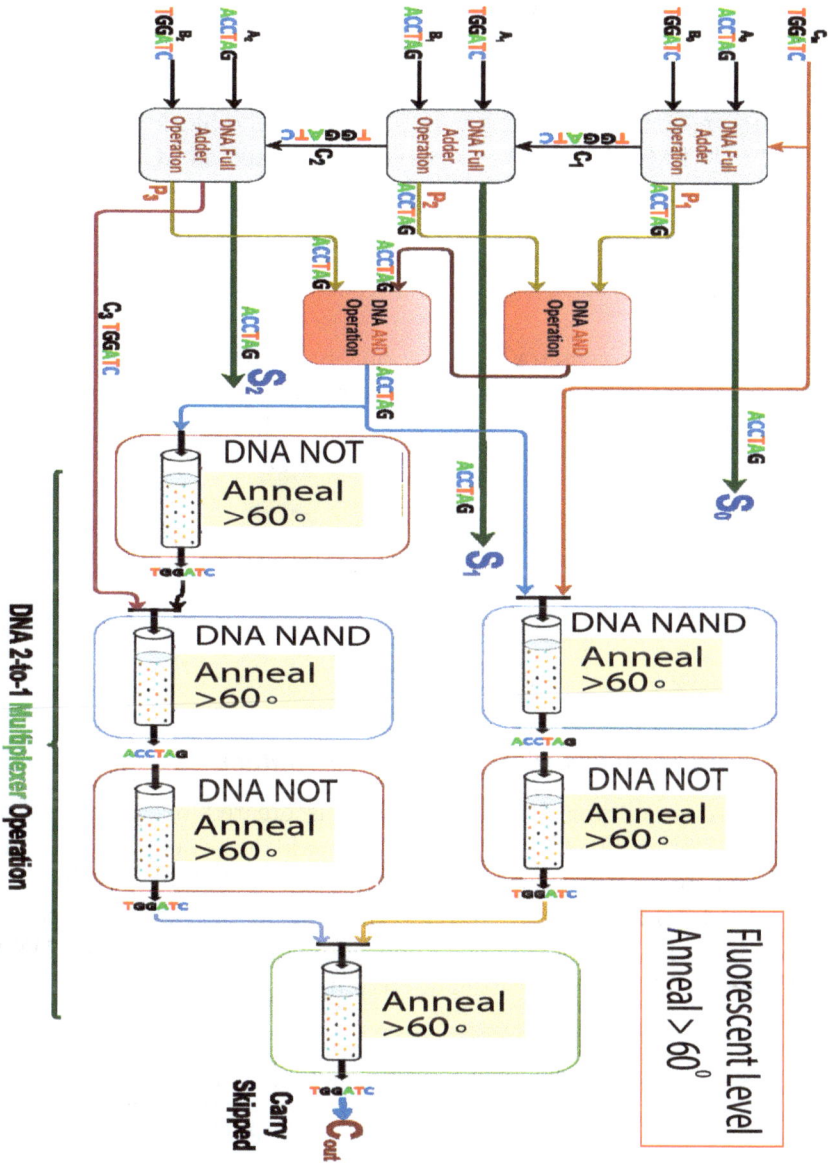

Figure 3.18.   General architecture of a 3-molecular DNA carry-skip adder.

3. For the third DNA full adder operation, input DNA sequences are
   $A_2 =$ ACCTAG and $B_2 =$ TGGATC. Therefore, $P_2 =$ ACCTAG.

Now it is easy to find all the values of P_i. So the block propagate is
BP = ACCTAG. Therefore, no carry is generated as ACCTAG. And
the sum of the operations is the values of all P_i's for each A_i and B_i.

3.8   DNA Half Subtractor

The DNA half subtractor has two output states: "D" and "Bout."
The sequence ACCTAG represents the true value "true" and the
sequence **TGGATC** represents the false value "false." The truth
table for a DNA half subtractor is given in Table 3.3.
    From Table 3.3,

$$\mathbf{D} = A'\,B + A\,B'$$

$$= A\ XOR\ B$$

$$Bout = A'\,B.$$

3.8.1   *Block diagram of a DNA half adder*

The DNA circuit for a half subtractor is created by a number of
DNA operations. To design a DNA half subtractor, one DNA XOR
operation, one DNA NOT operation, and one AND operation are
required. In this DNA circuit, one DNA XOR operation is needed
to produce the difference of the half adder. Additionally, one DNA
NOT and DNA AND operation are required to generate the borrowed

Table 3.3.   Truth table for a DNA half subtractor.

Inputs		Outputs	
A	B	D	Bout
TGGATC	TGGATC	TGGATC	TGGATC
TGGATC	ACCTAG	ACCTAG	ACCTAG
ACCTAG	TGGATC	ACCTAG	TGGATC
ACCTAG	ACCTAG	TGGATC	TGGATC

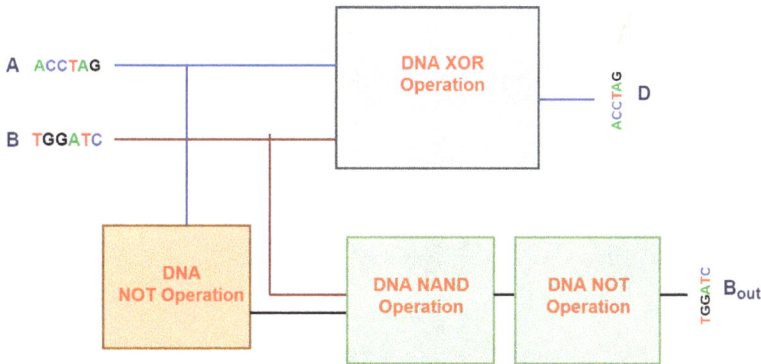

Figure 3.19. Block diagram of a DNA half subtractor.

molecular sequence. The block diagram of the DNA half subtractor is illustrated in Figure 3.19.

3.8.2 *Circuit architecture*

Figure 3.20 depicts the DNA circuit of the half subtractor. In this half subtractor circuit, the XOR operation for A and B inputs is done by a DNA operation and creates an output of D. As for the output of Bout, at first A is performed a DNA NOT operation and then both inputs A′ and B pass through a DNA AND operation.

3.8.3 *Working principle*

The DNA half adder needs two inputs, A and B. For inputs A and B, consider the following cases:

1. When both A and B are "true" or **ACCTAG**, the output sequences of both D and Bout are "false" or **TGGATC**.
2. When both A and B are "false" or **TGGATC**, the output sequences of both D and Bout are also "false" or **TGGATC**.
3. When A is "false" or **TGGATC** and B is "true" or **ACCTAG**, the output sequence of D is "true" or **ACCTAG** and Bout is "true" or **ACCTAG**.
4. When A is "true" or **ACCTAG** and B is "false" or **TGGATC**, the output sequence of D is "true" or **ACCTAG** and Bout is "false" or **TGGATC**.

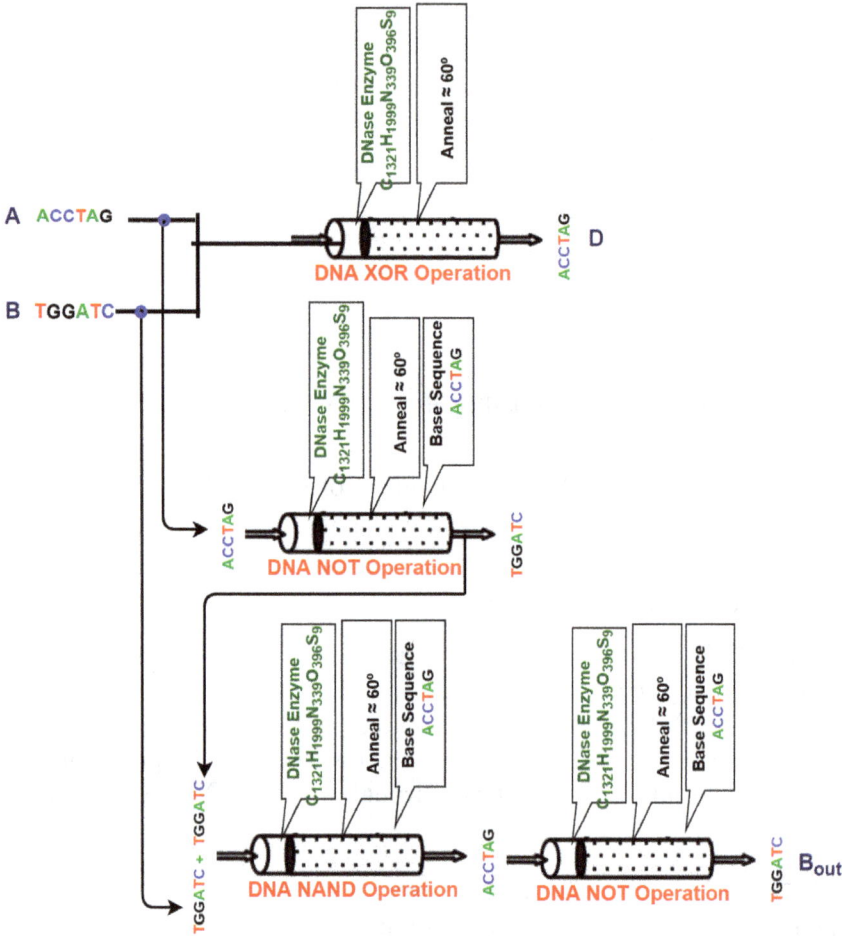

Figure 3.20.   DNA half subtractor circuit.

3.9   DNA Full Subtractor

A full subtractor is a combinational circuit that is developed to overcome the drawback of the half subtractor circuit. It can take three inputs; and after subtracting them, it creates two outputs. The sequence ACCTAG presents the true value "true" and the sequence **TGGATC** presents the false value "false." The truth table of a DNA full subtractor is shown in Table 3.4.

Table 3.4. Truth table of a DNA full subtractor.

Inputs			Outputs	
A	**B**	**Bin**	**D**	**Bout**
TGGATC	TGGATC	TGGATC	TGGATC	TGGATC
TGGATC	TGGATC	ACCTAG	ACCTAG	ACCTAG
TGGATC	ACCTAG	TGGATC	ACCTAG	ACCTAG
TGGATC	ACCTAG	ACCTAG	TGGATC	ACCTAG
ACCTAG	TGGATC	TGGATC	ACCTAG	TGGATC
ACCTAG	TGGATC	ACCTAG	TGGATC	TGGATC
ACCTAG	ACCTAG	TGGATC	TGGATC	TGGATC
ACCTAG	ACCTAG	ACCTAG	ACCTAG	ACCTAG

From Table 3.4,

$$\mathbf{D} = A'B'Bin + A'BBin' + AB'Bin' + ABBin$$
$$= Bin\,(A'B' + AB) + Bin'\,(AB' + A'B)$$
$$= Bin\,(A\ XNOR\ B) + Bin'\,(A\ XOR\ B)$$
$$= Bin\,(A\ XOR\ B)' + Bin'\,(A\ XOR\ B)$$
$$= Bin\ XOR\,(A\ XOR\ B)$$
$$= (A\ XOR\ B)\ XOR\ Bin;\ and$$
$$Bout = A'B'Bin + A'BBin' + A'BBin + ABBin$$
$$= Bin\,(AB + A'B') + A'B\,(Bin + Bin')$$
$$= Bin\,(A\ XNOR\ B) + A'B$$
$$= Bin\,(A\ XOR\ B)' + A'B.$$

3.9.1 *Block diagram of a DNA full subtractor*

The DNA circuit for a full subtractor is created by a number of DNA operations. To design a DNA full subtractor, two XOR, two AND, two NOT and an OR operations are required. In this DNA circuit, two DNA XOR operations are required to produce the difference (D) of the DNA full subtractor, and for Bout, two DNA AND operations, two DNA NOT, and one DNA OR operations are used along with the

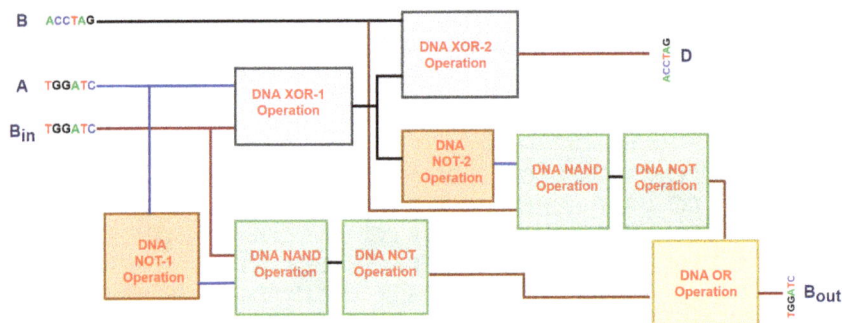

Figure 3.21.   Block diagram of a DNA full subtractor.

DNA XOR operation. The block diagram of the DNA full subtractor is illustrated in Figure 3.21.

3.9.2   *Circuit architecture of a DNA full subtractor*

Figure 3.22 shows the DNA circuit of the full subtractor. In this full subtractor, the first XOR operation for A and B inputs is done by a DNA XOR operation to create the output, A $\oplus$ B. This output, A $\oplus$ B, is used as one of the inputs of the second XOR operation and another input molecular sequence comes from bin. Finally, the output, D of the full subtractor is generated through the second DNA XOR operation which represents the difference of the full subtractor.

In addition, two DNA AND operations are needed to get another output sequence of Bout. At first AND operation occurs between the B input and the negation of A input to produce A'. B. As for the second AND operation, one input comes from the NOT operation of A $\oplus$ B, and another input is directly provided from Bin. These two input sequences then generate (A $\oplus$ B)'. Bin. Finally, the outputs of both DNA AND operations pass through a DNA OR operation to produce the value of Bout.

3.9.3   *Working principle of a DNA full subtractor*

The DNA full subtractor needs three inputs, A, B, and Bin and provides two outputs D and Bout. To understand the working principle of a DNA full subtractor, consider the values of inputs A, B,

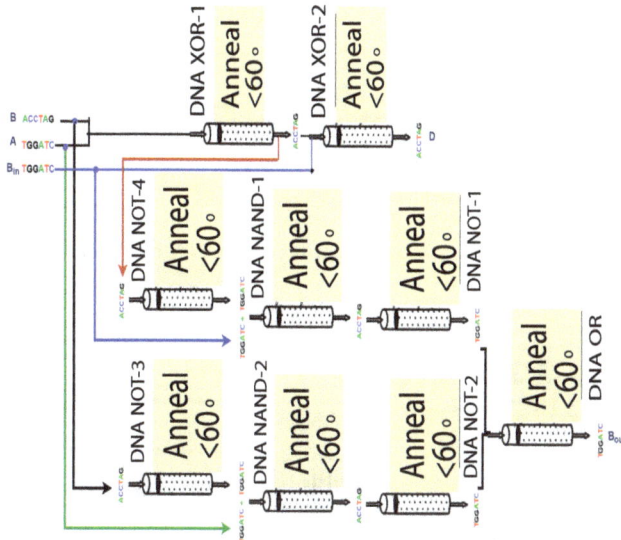

Figure 3.22. DNA full subtractor circuit.

and Bin which are **ACCTAG, TGGATC**, and **ACCTAG**, respectively. Then

1. The output of DNA XOR-1 is "true" or **ACCTAG,** the output of DNA AND-1 is "false" or **TGGATC**, and the output of DNA AND-2 is "false" or **TGGATC**.
2. Finally, the value of the output D is "false" or **TGGATC** after performing DNA XOR-2 operation; and Bout is also "false" or **TGGATC**.

3.10 DNA Multiplier

A DNA multiplier is a DNA combinational logic circuit that multiplies two DNA sequences. The two numbers are called multiplicand and multiplier, respectively and the outcome is called a product.

3.10.1 *Block diagram of a DNA multiplier*

A 2×2 DNA multiplier takes two 2-molecular input values and produces the product value of them. Figure 3.23 shows the block diagram

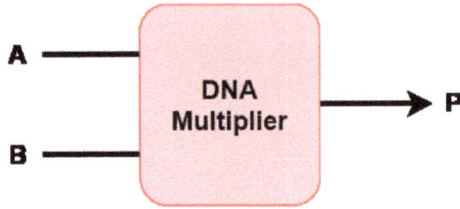

Figure 3.23.   Block diagram of a DNA multiplier.

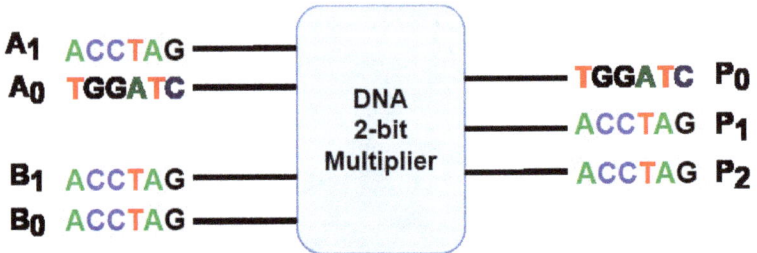

Figure 3.24.   Block diagram of a 2 × 2 DNA multiplier.

of the DNA multiplier, whereas Figure 3.24 shows the block diagram of the DNA 2 × 2 multiplier.

Suppose, multiplication is done within two 2-molecular inputs A and B.

Here, A = A_1A_0, and B = B_1B_0. So,

$$\textbf{A: } A_1 \quad A_0 \text{ (multiplicand)}$$
$$\textbf{B: } B_1 \quad B_0 \text{ (multiplier)},$$

$$\textbf{DNA-AND } A_1.B_0 \quad A_0.B_0$$
$$A_1.B_1 \quad A_0.B_1 \quad \textbf{X}$$
$$\textbf{DNA-OR } C_2 \ A_1.B_1 \ A_1.B_0 A_0.B_0$$
$$+ \ C_1 + A_0.B_1$$
$$\textbf{P: } \quad P_3 \ P_2 \ P_1 \ P_0.$$

The above calculation shows how a 2 × 2 molecular DNA multiplication is performed, where a 2 × 2 DNA multiplier produces a product of P = $P_3P_2P_1P_0$ using two 2-molecular inputs A = A_1A_0

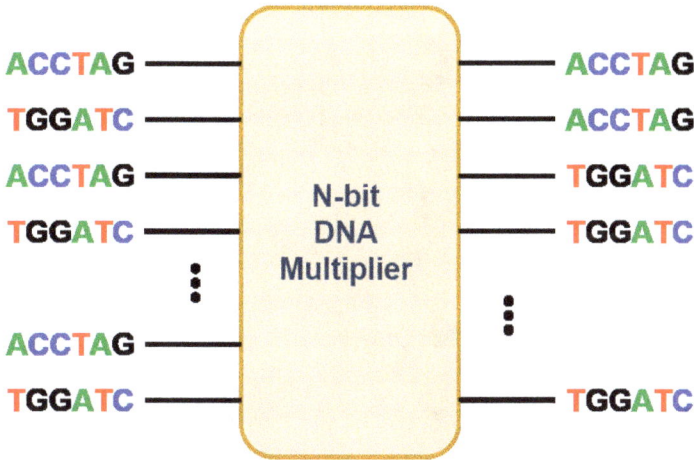

Figure 3.25. Block diagram of an n-molecular DNA multiplier.

and $B = B_1B_0$. At first, the DNA AND operation is performed as shown in the calculation. Then, the expected result is obtained using DNA AND operations. Figure 3.25 shows the block diagram of an n-molecular DNA multiplier.

3.10.2 *Circuit architecture of a DNA multiplier*

To construct the circuit diagram of the DNA 2×2 multiplier, at first, the circuit of the DNA AND operation is needed to perform DNA AND operations among them as shown in the calculation. Then, the DNA half-adders are used to perform the addition operations.

And while performing the addition process, a carry molecular sequence is generated which has to propagate the carry to the following add operations. However, the final carry molecular sequence is the most significant part of the output value.

Figure 3.26 shows the circuit architecture of a 2×2 DNA multiplier.

The architecture in Figure 3.26 consists of four DNA AND operational circuits (four DNA NANDs followed by four DNA NOT operations), two DNA OR, and two DNA XOR operational circuits. And they are combined in a way so that they can produce the expected output, which is the product of the two input molecular sequences.

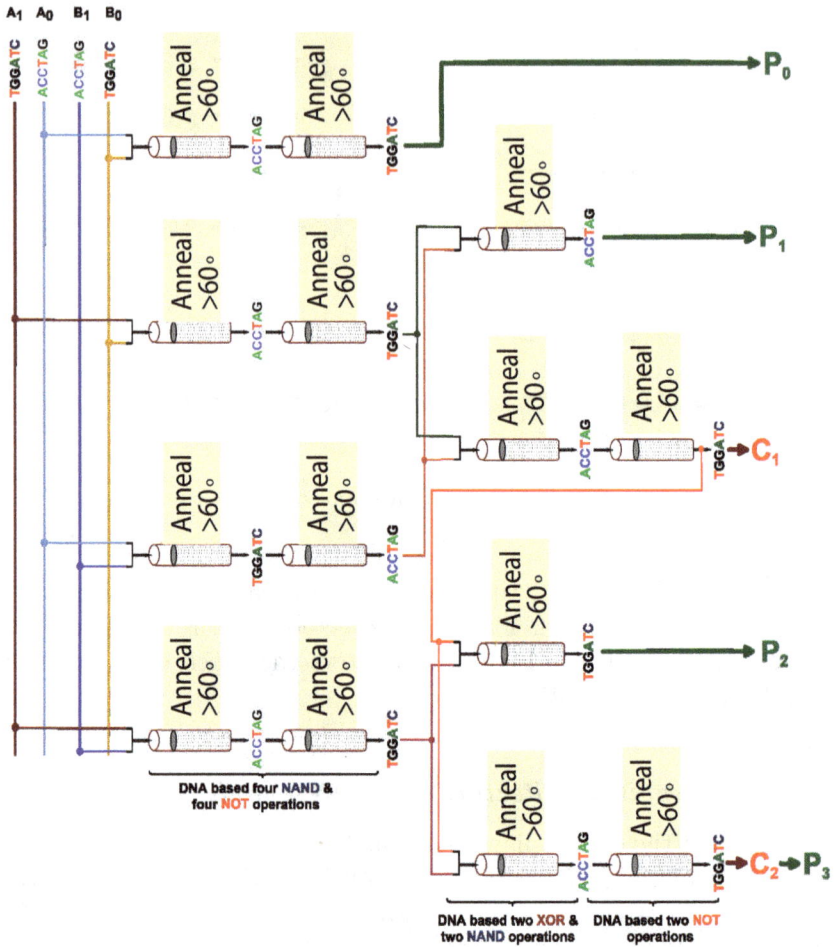

Figure 3.26.   Circuit architecture of a DNA 2 × 2 multiplier.

3.10.3   *Working principles of a 2 × 2 DNA multiplier circuit*

It is already described how the DNA 2 × 2 multiplier works. Now, let's describe the operational behavior of the designed architecture for the DNA 2 × 2 multiplier as shown in Figure 3.26.

Suppose, for multiplying 2 input values, where $A = 11$ and $B = 10$. So, the values of $A_0 = $ ACCTAG and $A_1 = $ ACCTAG. And the values of $B_0 = $ TGGATC and $B_1 = $ ACCTAG. Now,

1. The first DNA AND operation (includes DNA NAND followed by DNA NOT) receives input $A_0 = $ ACCTAG and $B_0 = $ TGGATC, which produces output TGGATC; and that is the value of P_0.
2. The second DNA AND operation receives input $A_1 = $ ACCTAG and $B_0 = $ TGGATC, which produces output TGGATC. And this DNA sequence works as an input DNA sequence of the first DNA XOR operation and the fifth DNA AND operation.
3. The third DNA AND operation receives input $A_0 = $ ACCTAG and $B_1 = $ ACCTAG, which produces output ACCTAG. And this output sequence works as an input DNA sequence of the first DNA XOR operation and the fifth DNA AND operation.
4. The fourth DNA AND operation receives input $A_1 = $ ACCTAG and $B_1 = $ ACCTAG, which produces output ACCTAG. And this DNA sequence works as an input DNA sequence of the second DNA XOR operation and the sixth DNA AND operation.
5. Now, the first DNA XOR operation receives input DNA sequences TGGATC (from Step 2) and ACCTAG (from Step 3), which produce ACCTAG as output. Therefore, P_1 is ACCTAG.
6. The fifth DNA AND operation receives input DNA sequences TGGATC (from Step 2) and ACCTAG (from Step 3), which produce TGGATC as output. And this output is the carry output DNA sequence, that works as an input for the next DNA XOR and DNA AND operations.
7. Now, the second DNA XOR operation receives input DNA sequences ACCTAG (from Step 4) and TGGATC (from Step 6). Thus, it produces ACCTAG as output. Therefore, P_2 is ACCTAG.
8. Finally, the sixth DNA AND operation receives input DNA sequences ACCTAG (from Step 4) and TGGATC (from Step 6), which produce TGGATC as output. And this output value is the value of P_3. Therefore, the most significant DNA sequence of the product result is TGGATC.

Consequently, the product of $A = 11$ and $B = 10$ is $P = P_3P_2P_1P_0 = 0110$, where ACCTAG = "1" and TGGATC = "0."

Consider some other 2-molecular values for A and B and describe the operational behavior of the circuit for the corresponding input.

3.11 DNA Divider

DNA division is an essential but frequently overlooked aspect of DNA arithmetic. Although the DNA division is not very difficult, it can be more difficult to grasp at first than the other DNA arithmetic operations. This is due to the fact that all other DNA arithmetic operations are comparable, yet DNA division is a sequence of an oddity. Besides, the architecture is much more complex than the others. And there are no specific rules that must be followed when performing DNA division. Furthermore, the circuit structure is significantly more complicated than the others.

3.11.1 *Block diagram of a DNA divider*

At first, the divisor value is made that is equivalent to the dividend value by adding the corresponding number of sequence TGGATC (i.e. zero) to the most significant sequence in divisor input.

Then, the DNA divider is designed for the 2-molecular dividend as shown in Figure 3.27. For the 4-molecular dividend, the n-molecular dividend and every pattern of dividend values, it is easy to construct the DNA divider circuit.

Figure 3.27 shows the general architecture of a 2-molecular DNA divider. Figure 3.28 presents the general architecture of a 4-molecular DNA divider. Figure 3.29 depicts the general architecture of an n-molecular DNA divider (here n DNA sequences mean the dividend value of the operation which is the size of n-molecular sequences).

It also needs n numbers of DNA NOT operations, n^2 numbers of DNA full subtractors, and $[n(n-1)+1]$ numbers of DNA 2-to-1 multiplexers to construct the DNA divider which performs the division operation of n-molecular inputs.

3.11.2 *Circuit architecture of a DNA divider*

The main components to construct the architecture of a 2-molecular DNA divider are four DNA full subtractors, where first two subtractors are connected sequentially with the inputs B_0, and A_1, and B_1,

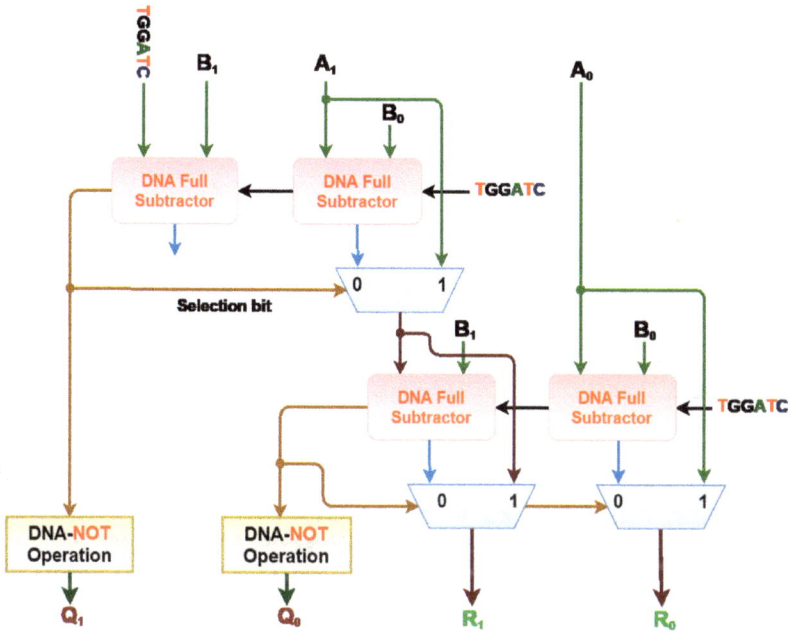

Figure 3.27.   General architecture of the 2-molecular DNA divider.

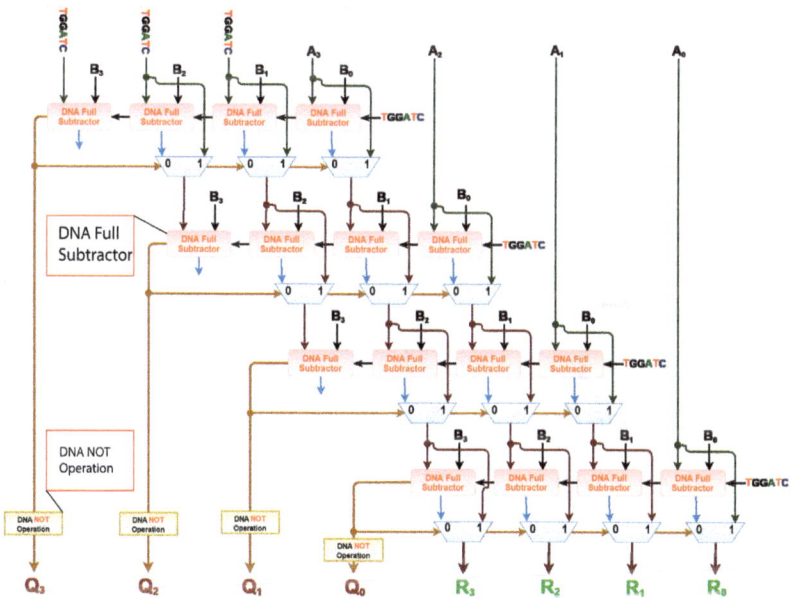

Figure 3.28.   General architecture of the 4-molecular DNA divider.

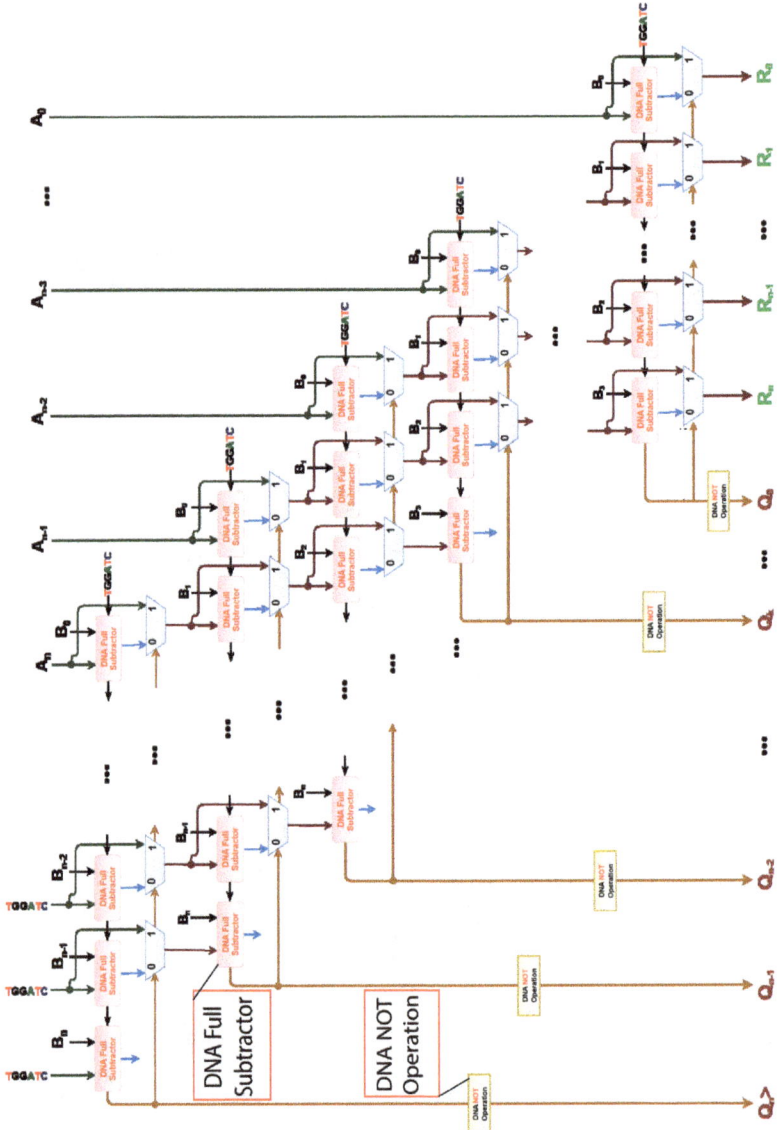

Figure 3.29.   General architecture of the *n*-molecular DNA divider.

and TGGATC (i.e. 0). The output from second DNA full-subtractors is the selection DNA sequence of the first DNA multiplexer; and the DNA NOT value of that output is the output of Q_1.

The next two subtractors are connected sequentially with the inputs B_0 and A_0, and B_1, and the output of the first multiplexer. The output of the fourth full subtractor is the selection DNA sequence for the next two DNA multiplexers, and the DNA NOT value of that output is the output of Q_0. And finally, the last two multiplexer's outputs are the value of the remainders as shown in Figure 3.30.

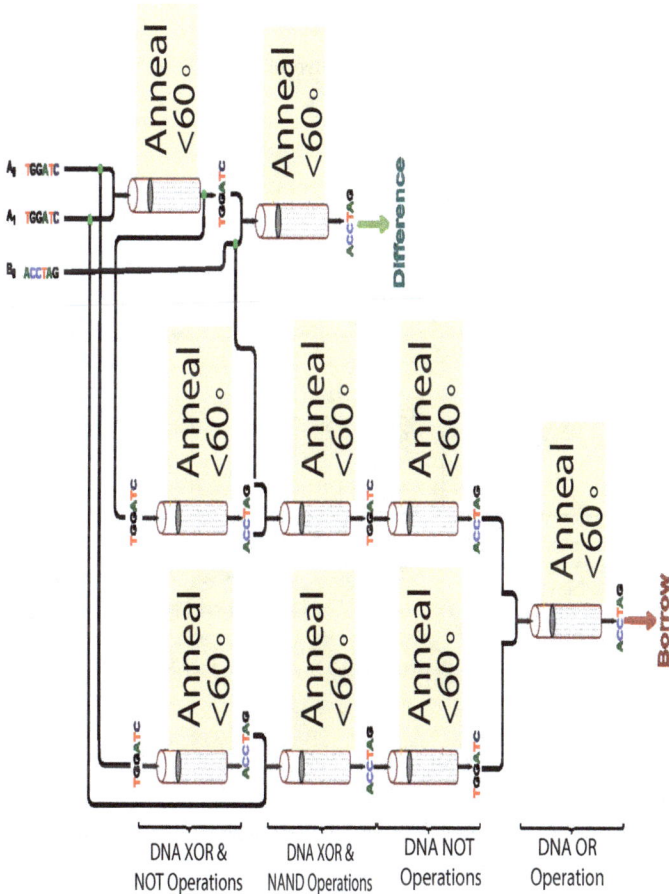

Figure 3.30. Circuit architecture of a DNA full subtractor.

TGGATC

DNase Enzyme
$C_{1321}H_{1999}N_{339}O_{396}S_9$

Anneal < 60°

Base Sequence
ACCTAG

ACCTAG

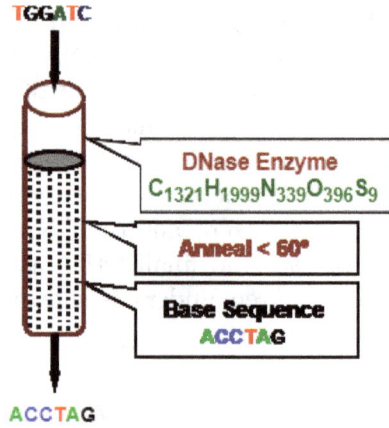

Figure 3.31.   Circuit architecture of DNA NOT operation.

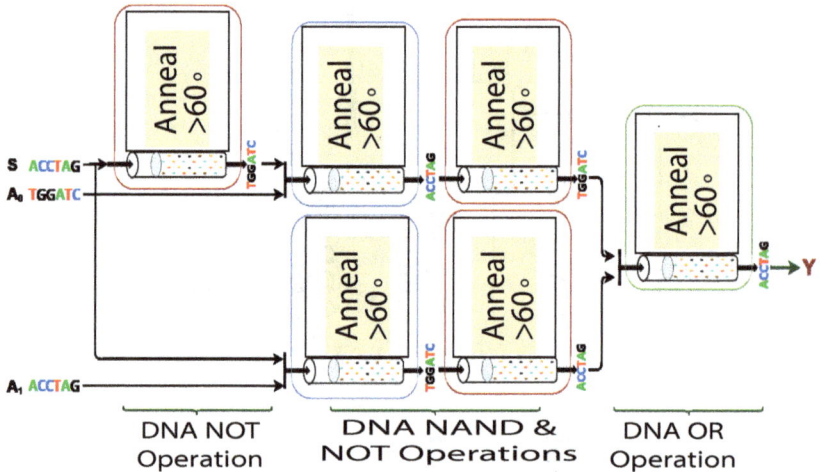

Figure 3.32.   Circuit architecture of 2-to-1 DNA multiplexer.

So, the DNA full subtractor, DNA NOT operation, and 2-to-1 DNA multiplexer are needed to construct the circuit architecture of 2-molecular DNA divider. Figure 3.30 shows the circuit architecture of DNA full subtractor, Figure 3.31 presents the circuit architecture of DNA NOT operation and Figure 3.32 depicts the circuit architecture of 2-to-1 DNA multiplexer.

3.11.3 *Working principles of a DNA divider*

Suppose, the 2-molecular dividend is A and the 2-molecular divisor is B, where, A = 11 and B = 10. Therefore, A_0 = ACCTAG, and A_1 = ACCTAG. And B_0 = TGGATC and B_1 = ACCTAG. These inputs generate the quotient Q = Q_1Q_0, and the remainder as R = R_1R_0.

The behavior of the DNA 2-molecular divider circuit for these molecular sequences is discussed as follows:

1. The operation of the DNA full subtractor starts from the right side, and the operation of the DNA multiplexer begins from the left side of the figure as shown in Figure 3.30. The first DNA full subtractor gets inputs B_0 = TGGATC and A_1 = ACCTAG. And the other input of the full subtractor is TGGATC. Therefore, the difference output is ACCTAG, which works as the input of the first DNA multiplexer. And the borrow output will be TGGATC which will work as an input of the second DNA full-subtractor.
2. The second DNA full subtractor gets inputs B_1 = ACCTAG and TGGATC from Step 1, and the other input of the full subtractor is TGGATC. Therefore, the borrow output is TGGATC which works as the selection input of the first DNA multiplexer, and the DNA NOT operation of this borrow, DNA sequence ACCTAG is the value of Q_1. So, Q_1 = TGGATC.
3. Now, consider the first DNA multiplexer operation. It gets the selection input ACCTAG (from Step 2), and also ACCTAG (from Step 1), and the value of A1 = ACCTAG. Therefore, the output is ACCTAG. And this output works as an input of the fourth DNA full subtractor and the second DNA multiplexer.
4. The third DNA full subtractor gets inputs B_0 = TGGATC and A_0 = ACCTAG. And the other two inputs of the full subtractor are TGGATC. Therefore, the difference output is ACCTAG, which works as the input of the third DNA multiplexer. The borrow output sequence is TGGATC which works as an input of the fourth DNA full subtractor.
5. The fourth DNA full subtractor gets inputs B_1 = ACCTAG and TGGATC from Step 4 and ACCTAG from Step 3. Therefore, the borrow output sequence is TGGATC which works as the selection input of the second and third DNA multiplexer, and the DNA NOT operation of this borrow output sequence TGGATC is the

value of Q_0. So, $Q_0 = $ ACCTAG. And the difference output is also TGGATC that works as an input of the second DNA multiplexer.

6. The quotient part is completed. The second DNA multiplexer gets the selection input TGGATC (from Step 5), another input TGGATC from Step 5, and ACCTAG from Step 3. Thus, the output is TGGATC and the value of R_1 is TGGATC.

7. Finally, the third DNA multiplexer gets the selection input TGGATC (from Step 5), another input ACCTAG from Step 3, and ACCTAG from the value of A_0. Therefore, the output is ACCTAG and the value of R_0 is ACCTAG.

Thus, after performing the division operation between the dividend $A = 11$ and the divisor $B = 10$, the quotient $Q = 01$ and the remainder $R = 01$.

3.12   DNA Comparator

A comparator that compares two input signals and gives output as which input is larger or smaller or equal. The 1-molecular DNA sequence comparator takes two single DNA sequence inputs, A and B. The values of the inputs can be **TGGATC** and **ACCTAG**. This circuit gives three outputs X, Y, and Z, where X is true when A is smaller than B, Y is true when A is equal to B, and Z is true when A is greater than B. Here, Table 3.5 shows the truth table of the 1-molecular DNA comparator.

Table 3.5.   1-Molecular DNA comparator truth table.

Inputs		Outputs		
A	**B**	**X**	**Y**	**Z**
		$A < B$	$A = B$	$A > B$
TGGATC	TGGATC	TGGATC	ACCTAG	TGGATC
TGGATC	ACCTAG	ACCTAG	TGGATC	TGGATC
ACCTAG	TGGATC	TGGATC	TGGATC	ACCTAG
ACCTAG	ACCTAG	TGGATC	ACCTAG	TGGATC

From Table 3.5, X, Y, and Z are as follows:

$$\mathbf{X} = A^0.B^1,$$

$$\mathbf{Y} = A^0.B^0 + A^1.B^1 = A \text{ XNOR } B,$$

$$\mathbf{Z} = A^1.B^0.$$

3.12.1 *Block diagram of a 1-molecular DNA comparator*

Now, using the equations, the DNA 1-molecular comparator is designed. Figure 3.33 displays the block diagram of the DNA 1-molecular comparator. Here two inputs are A and B, and three outputs are X, Y, and Z.

3.12.2 *Circuit architecture of a 1-molecular DNA comparator*

DNA 1-molecular comparator needs two DNA AND operations: two DNA NOT operations and one DNA XNOR operation as shown in Figure 3.34. DNA AND-1 is connected to A′ and B. It performs A′. B which is the result of Z. In the same way, DNA AND-2 provides the output of X. To get the value of Y, both the outputs of AND operation act as the inputs of the DNA XNOR operation.

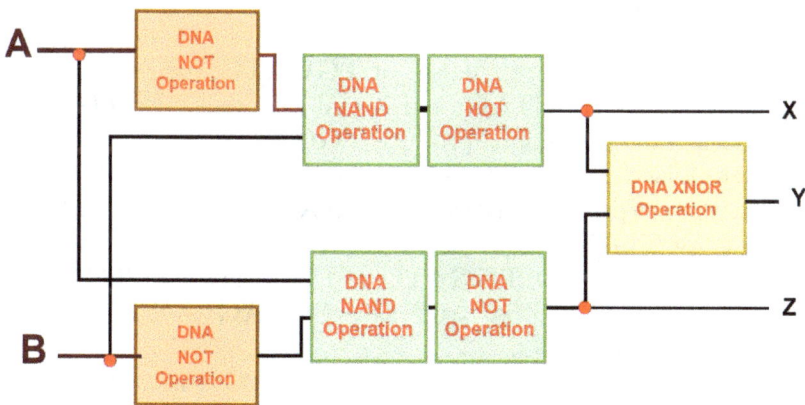

Figure 3.33. Block diagram of 1-molecular DNA comparator.

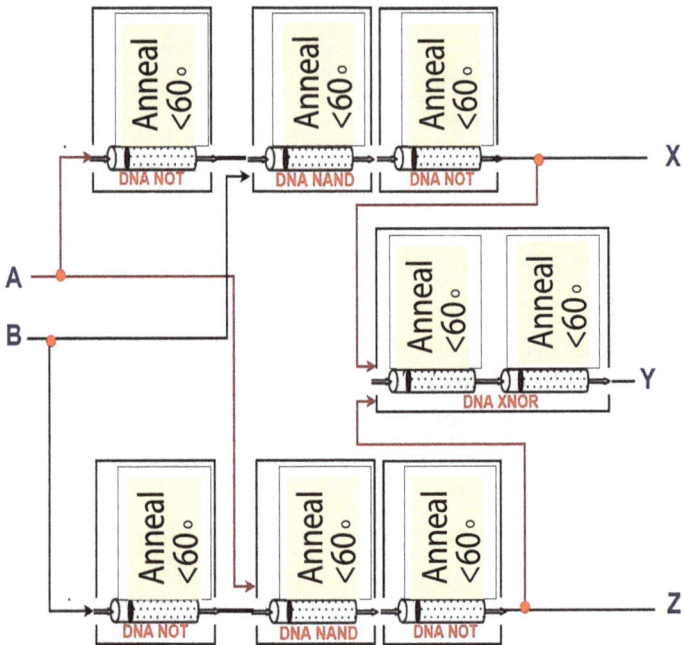

Figure 3.34.   1-Molecular DNA comparator circuit.

3.12.3   *Working principle of a 1-molecular DNA comparator*

The DNA comparator needs two inputs, A and B. For various inputs of A and B, consider the following cases:

1. When A and B are both "false" or **TGGATC**, then the output of Y is "true" or **ACCTAG**; and X and Z is "false" or **TGGATC** as both AND operations produce **TGGATC**.
2. When A is "false" or **TGGATC** and B is "true" or **ACCTAG**, then the output of X is "true" or **ACCTAG**; and Y and Z are "false" or **TGGATC**.
3. When A is "true" or **TGGATC** and B is "false" or **TGGATC**, then the output of Z is "true" or **ACCTAG**; and X and Y are "false" or **TGGATC**.
4. When both A and B are "true" or **ACCTAG**, then the output of Y is "true" or **ACCTAG**; and X and Z are "false" or **TGGATC** as both AND operations produce **TGGATC**.

3.13   Summary

DNA carry ripple adder is the chain of DNA full adders, where each DNA full adder's output carry sequence is connected to the carry input of the next higher-order DNA full adder in the chain. An n-molecular adder requires n DNA full adders to operate. A 4-molecular DNA parallel adder requires a total execution time of 40 hours. And the n-molecular DNA adder requires $(8n + 8)$ hours to execute. The DNA carry-lookahead adder determines the carry molecular sequences before the sum, which minimizes the waiting time to calculate the result of the larger value sequences of the adder. To perform a 3-molecular addition, the DNA carry-lookahead adder takes only 16 hours. Thus, the DNA carry-lookahead adder is the fastest adder. The DNA carry-skip adder skips carry molecular sequences when it finds that there is no carry generating which is equivalent to TGGATC. It performs well in the best cases.

Bibliography

Rana Barua and Janardan Misra. Binary arithmetic for DNA computers. In *International Workshop on DNA-Based Computers*, pp. 124–132. Springer, Japan, 2002.

Akihiro Fujiwara, Ken'Ichi Matsumoto, and Wei Chen. Procedures for logic and arithmetic operations with DNA molecules. *International Journal of Foundations of Computer Science*, 15(03): 461–474, 2004.

Vineet Gupta, Srinivasan Parthasarathy, and Mohammed Zaki. Arithmetic and logic operations with DNA. USA, 1999, doi: 10.1090/dimacs/048/11.

V. Sudha and K. S. Easwarakumar. A comprehensive study on the arithmetic operations in DNA computing. In *Computational Vision and Bio-Inspired Computing*, pp. 191–198. Springer, Singapore, 2021.

Chapter 4

DNA Combinational Circuits

4.1 Introduction

A DNA computer is essentially a collection of specially selected DNA strands, the combinations of which will use biomolecular manipulation to solve computational problems while also exploring natural processes as computational models. Building algorithms for DNA computing requires a strong background in biology and computer science.

There are two types of DNA computing: theoretical and practical. Theoretical research began by observing the structure and dynamics of DNA and progressed to propose formal models of DNA computers for performing theoretical operations. The practical side of DNA computing has advanced at a much slower pace because laboratory work is time-consuming and subject to several constraints. The following are the key characteristics of DNA computing:

1. Dense data storage;
2. Massively parallel computation; and
3. Extraordinary energy efficiency.

This chapter discusses the combinational circuits in DNA computing. DNA multiplexer, DNA demultiplexer, DNA encoder, and DNA decoder will be described here with appropriate figures, algorithms, and their working principles.

4.2 Combinational DNA Circuits

A combinational DNA circuit is one kind of fundamental DNA circuits in which there are various DNA circuits, such as encoders, decoders, multiplexers, and demultiplexers. The following are some of the properties of combinational DNA circuits:

1. At any given point, the output of a combinational DNA circuit is solely determined by the levels present at the input terminals.
2. There is no memory in the combinational circuit. The prior state of the input has no bearing on the circuit's current state.
3. An n number of inputs and m number of outputs can be found in a combinational DNA circuit.

4.2.1 *DNA multiplexer*

A DNA multiplexer is a combinational circuit. A DNA *multiplexer* makes it possible for several input signals to share one device or resource.

4.2.1.1 *Block diagram of a DNA multiplexer*

There are only two inputs, I0 and I1, one selection line, S, and one output, Y in a 2-to-1 multiplexer.

The truth table of 2-to-1 DNA multiplexer is shown in Table 4.1 and the truth table of 4-to-1 DNA multiplexer is shown in Table 4.2.

From Table 4.1,

$$Y = S0'.I0 + S0.I1.$$

Table 4.1. Truth table of DNA 2-to-1 multiplexer.

Inputs	Outputs
S	Y
TGGATC	I0
ACCTAG	I1

Table 4.2.   Truth table of 4-to-1 DNA multiplexer.

Inputs		Outputs
S1	S0	Y
TGGATC	TGGATC	I0
TGGATC	ACCTAG	I1
ACCTAG	TGGATC	I2
ACCTAG	ACCTAG	I3

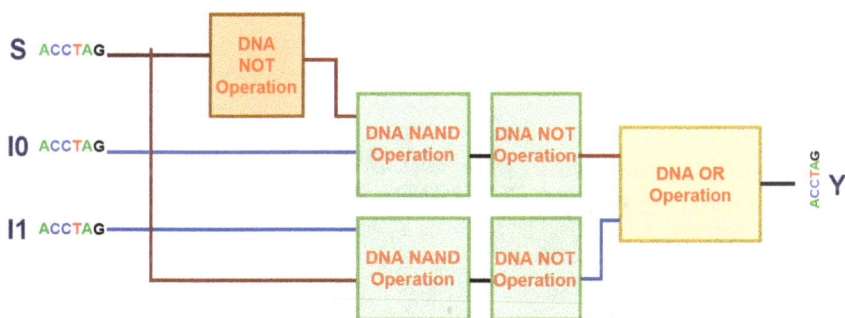

Figure 4.1.   Block diagram of 2-to-1 DNA multiplexer.

Figure 4.1 portrays the block diagram of a DNA 2-to-1 multiplexer. For this circuit, one DNA NOT operation, two DNA AND operations, and one DNA OR operation are needed.

From Table 4.2,

$$Y = S1'\ S0'\ I0 + S1'\ S0\ I1 + S1\ S0'\ I2 + S1\ S0\ I3.$$

Figure 4.2 displays the block diagram of a 4-to-1 DNA multiplexer. In this circuit, two DNA NOT operations, eight DNA AND operations, and three DNA OR operations are needed.

4.2.1.2   *Circuit architecture of a 2-to-1 DNA multiplexer*

Figure 4.3 depicts the DNA circuit of the 2-to-1 multiplexer. Firstly, each input of I0 and I1 is separately performed two DNA AND operations. The input of the selection line, S is connected directly to the

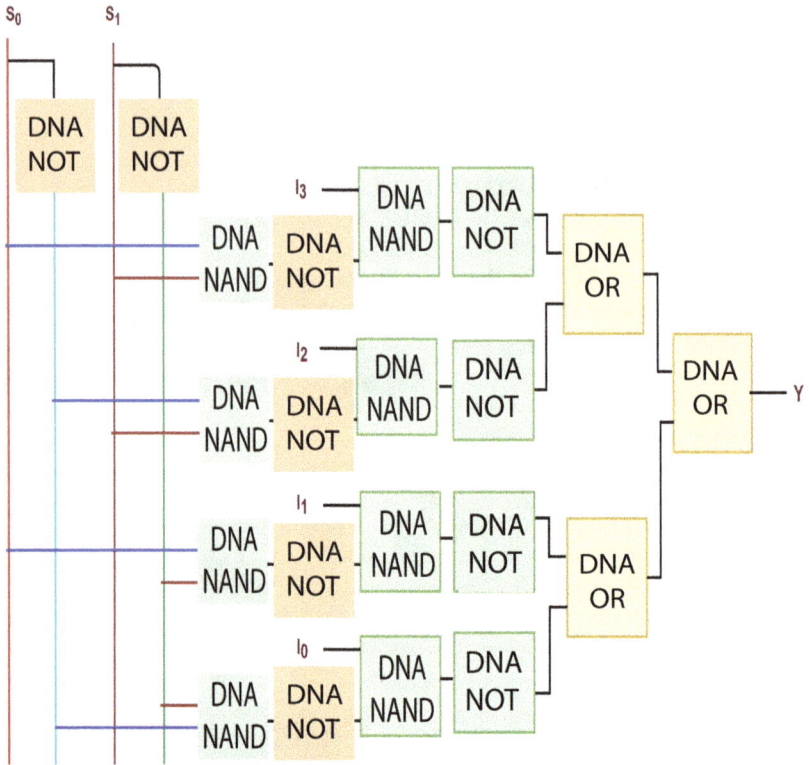

Figure 4.2.   Block diagram of a 4-to-1 DNA multiplexer.

AND-1 operation, whereas, another input is connected to the DNA AND-2 operation after completing the DNA NOT operation. Finally, both the outputs of DNA AND operations are passed through DNA OR operation generating the value of Y.

Figure 4.4 gives the view of the 4-to-1 DNA multiplexer circuit. First, each input of I0, I1, I2, and I3 separately performed four DNA AND operations. The other inputs of these AND operations come from the outputs of another four DNA AND operations which are executed by two selection lines, S0 and S1. Each DNA AND operation needs a DNA NAND and a DNA NOT operation to obtain its desired output. Finally, the outputs of DNA NOT-2, DNA NOT-4, DNA NOT-6, and DNA NOT-8 perform three DNA OR operations to obtain the value of Y.

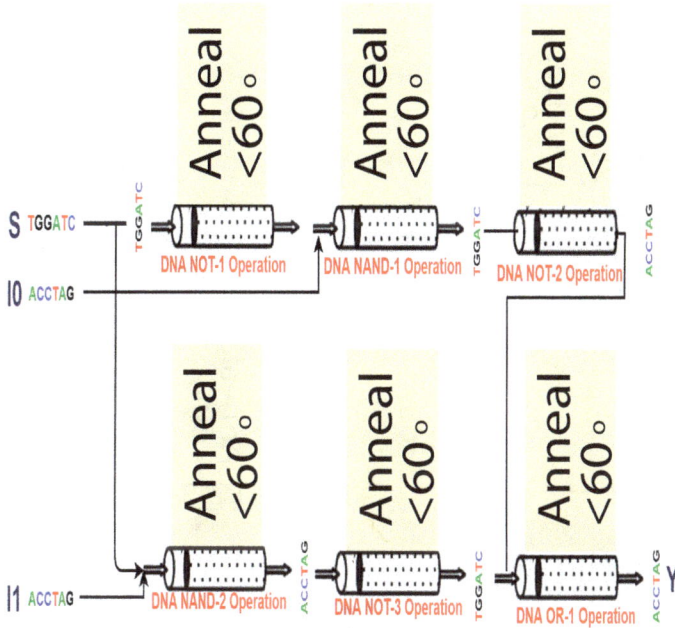

Figure 4.3. 2-to-1 DNA multiplexer circuit.

4.2.1.3 *Working principle of a DNA multiplexer*

DNA 2-to-1 Multiplexer

The DNA 2-to-1 multiplexer needs one selection line S0 and two inputs, I0 and I1. When inputs I0 and I1 are "true," or **ACCTAG**, consider the following cases:

(i) The output Y = I0, when the input sequence S is "false" or **TGGATC**; and

(ii) The output Y = I1, when the input sequence S is "true" or **ACCTAG**.

DNA 4-to-1 Multiplexer

The DNA 4-to-1 multiplexer needs two selection lines S0 and S1, and four inputs, I0-I3. When all inputs I0–I3 are "true," or **ACCTAG**, consider the following cases:

(i) The output Y = I0, when the input sequences S1 and S0 are "false" or **TGGATC**;

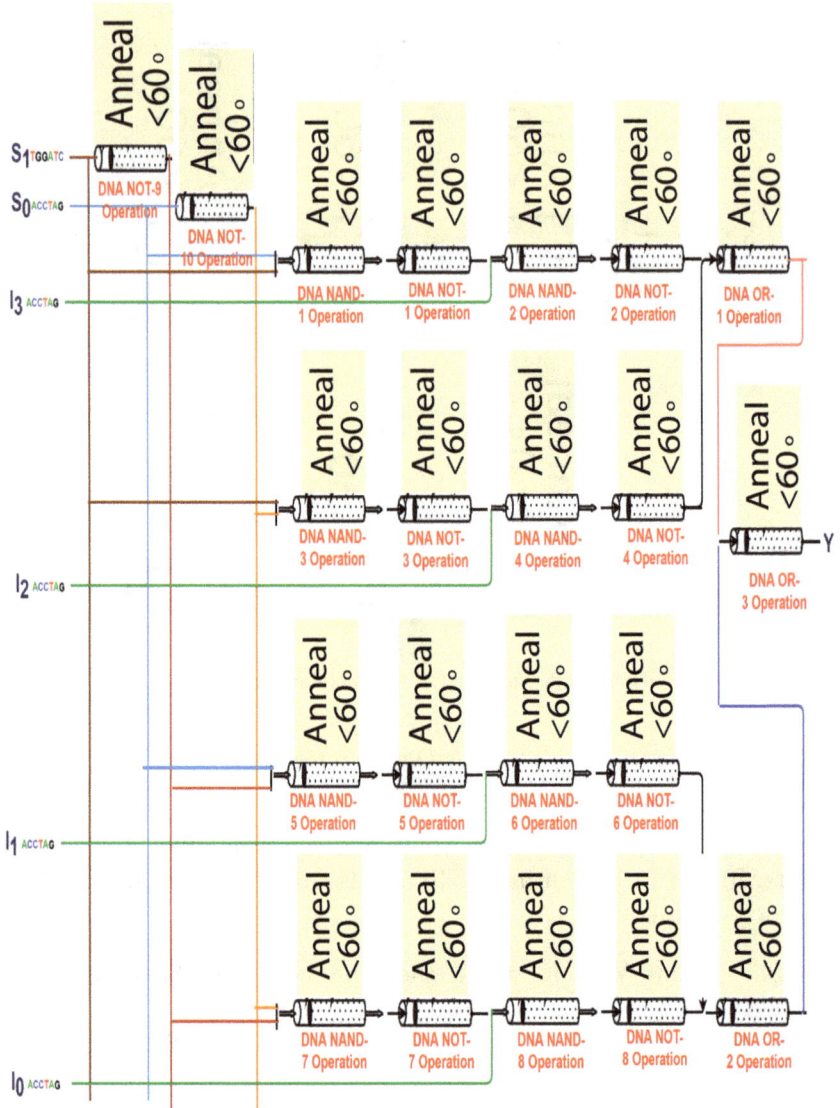

Figure 4.4.   4-to-1 DNA multiplexer circuit.

(ii) The output Y = I1, when the input S1 is "false" or **TGGATC** and input S0 is "true" or **ACCTAG**; and
(iii) The output Y = I2, when the input S0 is "false" or **TGGATC** and input S1 is "true" or **ACCTAG**.
(iv) The output Y = I3, when the inputs S0 and S1 are "true" or **ACCTAG**.

4.2.2 DNA demultiplexer

Demultiplexer is a data distributor which is read as DEMUX. It is quite opposite to multiplexer or MUX. It is a process of taking information from one input and transmitting over one of many outputs.

4.2.2.1 Block diagram of a 1-to-2 DNA demultiplexer

1-to-2 DNA Demultiplexer

There are only two inputs, I0 and I1, one selection line, S, and one output, Y in a 2-to-1 DNA multiplexer. Table 4.3 shows the truth table of 1-to-2 DNA demultiplexer.

From Table 4.3,

$$I0 = S'.D; \text{ and}$$
$$I1 = S.D.$$

Figure 4.5 presents the block diagram of a DNA 1-to-2 demultiplexer. For this circuit, one DNA NOT operation and two DNA AND operations are needed.

Figure 4.6 shows the block diagram of a 1-to-4 DNA demultiplexer. In this circuit, two DNA NOT and eight DNA AND operations are needed and Table 4.4 shows the truth table of a 1-to-4 DNA demultiplexer.

Table 4.3. Truth table of 1-to-2 DNA demultiplexer.

Inputs	Outputs	
S	I_0	I_1
TGGATC	TGGATC	D
ACCTAG	D	TGGATC

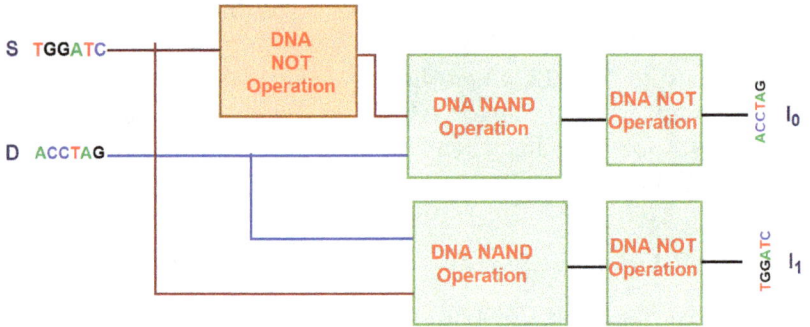

Figure 4.5.　Block diagram of a 1-to-2 DNA demultiplexer.

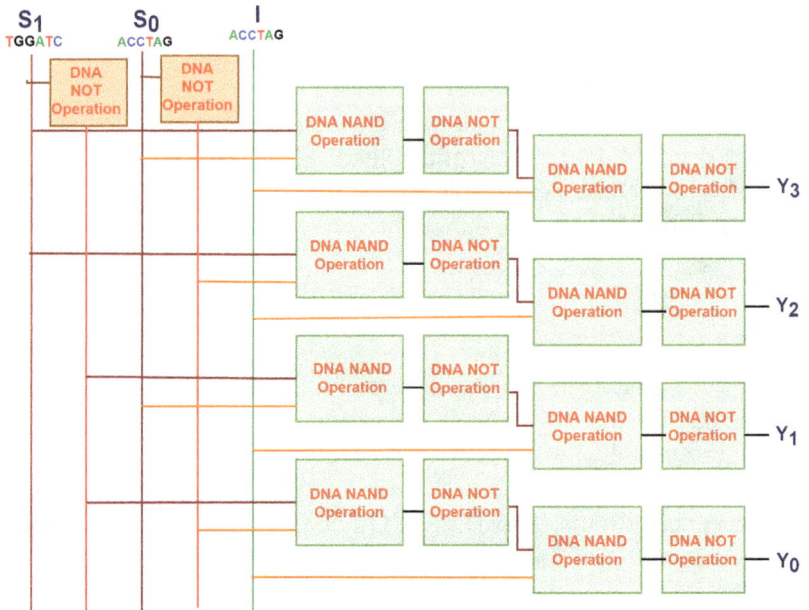

Figure 4.6.　Block diagram of a 1-to-4 DNA demultiplexer.

From Table 4.4,

$$Y0 = S1'\ S0'\ I;$$
$$y1 = S1'\ S0\ I;$$
$$y2 = S1\ S0'\ I;\ \text{and}$$
$$y3 = S1\ S0\ I.$$

Table 4.4.   Truth table of a 1-to-4 DNA demultiplexer.

Inputs		Outputs			
S1	S0	Y0	Y1	Y2	Y3
TGGATC	TGGATC	I		TGGATC	TGGATC	TGGATC
TGGATC	ACCTAG	TGGATC	I		TGGATC	TGGATC
ACCTAG	TGGATC	TGGATC	TGGATC	I		TGGATC
ACCTAG	ACCTAG	TGGATC	TGGATC	TGGATC	I

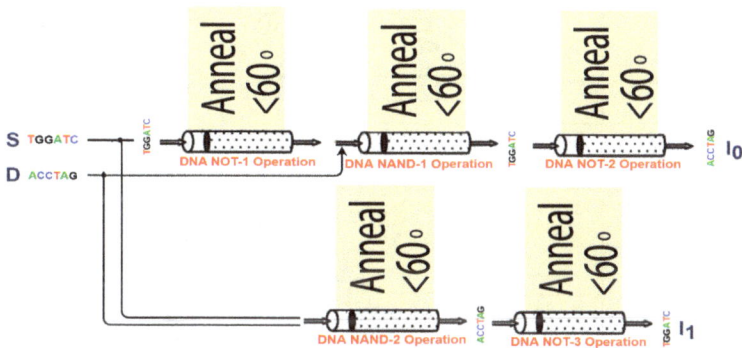

Figure 4.7.   1-to-2 DNA demultiplexer circuit.

4.2.2.2   *Circuit architecture of a DNA demultiplexer*

Figure 4.7 depicts the DNA circuit of the 1-to-2 demultiplexer. At first, input D connects with two DNA AND operations directly, whereas other inputs come from the selection input, S. Based on the values of S and D, only one DNA AND operation provides the output of **ACCTAG**.

Figure 4.8 gives the view of the 1-to-4 demultiplexer DNA circuit. First, the input I directly connects with four DNA AND operations. The other input of these AND operations comes from the output of another four DNA AND operations which are executed between two selection lines, S0 and S1. To obtain the desired output, each DNA AND operation requires a DNA NAND and a DNA NOT operation. Finally, the outputs of NOT-2, NOT-4, NOT-6, and NOT-8 generate the sequence of Y0, Y1, Y2, and Y3.

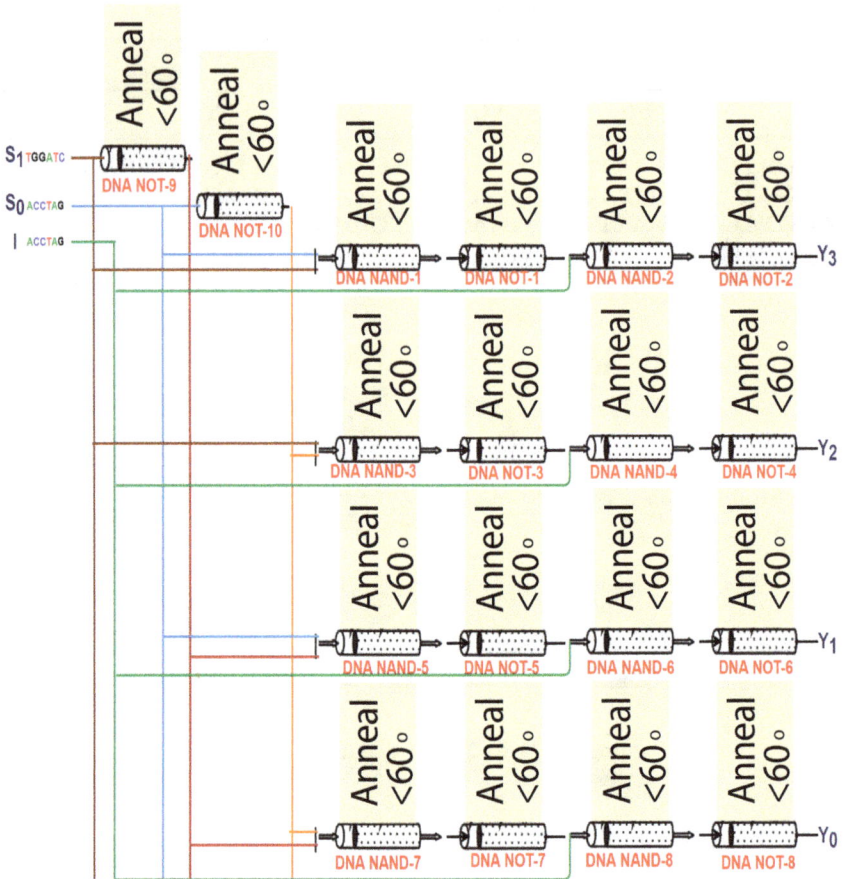

Figure 4.8.   1-to-4 DNA demultiplexer circuit.

4.2.2.3   *Working principle of a DNA demultiplexer*

DNA 1-to-2 Demultiplexer

The DNA 1-to-2 demultiplexer needs one selection line S0 and one input, D. When the input D is "true," or **ACCTAG**, consider the following cases:

(i) The output I0 = S′. D evaluates "true" or **ACCTAG**, when the input sequence S is "false" or **TGGATC**. Thus, I1 is "false" or **TGGATC**.

(ii) The output I1 = S. D evaluates "true" or **ACCTAG**, when the input sequence S is "true" or **ACCTAG**. So, I0 is "false" or **TGGATC**.

DNA 1-to-4 Demultiplexer

The DNA 1-to-4 demultiplexer needs two selection lines S0 and S1, and one input, I. When input I is "true" or **ACCTAG**, consider the following cases:

(i) The output Y0 = S0$'$. S1$'$. I evaluates "true" or **ACCTAG**, when the input sequences S1 and S0 are "false" or **TGGATC**. Thus, the output line Y0 is "true" or **ACCTAG** and Y1 to Y3 are "false" or **TGGATC**.

(ii) The output Y1 = S0.S1$'$. I evaluates "true" or **ACCTAG**, when the input S1 is "false" or TGGATC and input S0 is "true" or **ACCTAG**. So, the output line Y1 is "true" or **ACCTAG** and Y0, Y2, and Y3 are "false" or **TGGATC**.

(iii) The output Y2 = S0$'$.S1. I evaluates "true" or **ACCTAG**, when the input S0 is "false" or **TGGATC** and input S1 is "true" or **ACCTAG**. Thus, the output line Y2 is "true" or **ACCTAG** and Y0, Y1 and Y3 are "false" or **TGGATC**.

(iv) The output Y3 = S0.S1. I evaluates "true" or **ACCTAG**, when the inputs S0 and S1 are "true" or **ACCTAG**. So, the output line Y3 is "true" or **ACCTAG** and Y0 to Y2 are "false" or **TGGATC**.

4.2.3   *DNA encoder*

DNA encoder is a combinational circuit. In general, an encoder is a device or process that converts data from one format to another.

4.2.3.1   *Block diagram of a DNA encoder*

There are only two inputs, Y0 and Y1, and one output, A in a 2-to-1 DNA encoder. The truth table of a 2-to-1 DNA encoder is given in Table 4.5.

From Table 4.5,

$$A = Y0'.Y1.$$

Table 4.5. Truth table of 2-to-1 DNA encoder.

Inputs		Outputs
Y1	Y0	A
TGGATC	ACCTAG	TGGATC
ACCTAG	TGGATC	ACCTAG

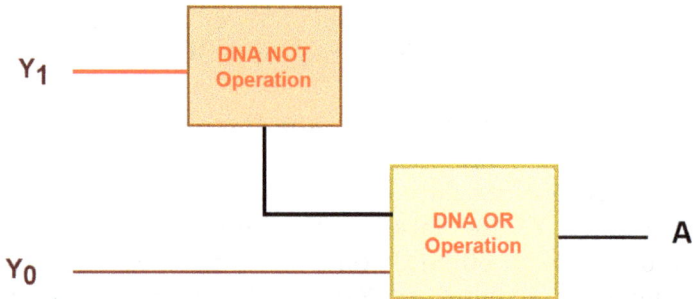

Figure 4.9. Block diagram of a 2-to-1 DNA encoder.

Table 4.6. Truth table of a 4-to-2 DNA encoder.

Inputs				Outputs	
Y0	Y1	Y2	Y3	A1	A0
I	TGGATC	TGGATC	TGGATC	TGGATC	TGGATC
TGGATC	I	TGGATC	TGGATC	TGGATC	ACCTAG
TGGATC	TGGATC	I	TGGATC	ACCTAG	TGGATC
TGGATC	TGGATC	TGGATC	I	ACCTAG	ACCTAG

Figure 4.9 presents the block diagram of a 2-to-1 DNA encoder. For this circuit, one DNA NOT operation and one DNA OR operation are needed.

4-to-2 DNA Encoder: There are only four inputs, Y0, Y1, Y3, Y4 and two outputs, A1 and A0 in a 4-to-2 DNA encoder. The truth table of a 4-to-2 DNA encoder is given in Table 4.6.

From Table 4.6,

$$A_1 = Y_3 + Y_2; \text{ and}$$

$$A_0 = Y_3 + Y_1.$$

Figure 4.10 shows the block diagram of a 4-to-2 DNA encoder. In this circuit, only two DNA OR operations are needed.

4.2.3.2 *Circuit architecture of a DNA encoder*

Figure 4.11 depicts the DNA circuit of the 2-to-1 encoder. Firstly, input Y0 performs a DNA NOT operation. After that, this output, along with Y1 input, transfers to a DNA OR operation to get the result of final output A_0.

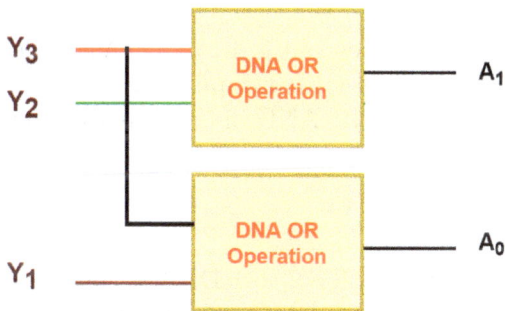

Figure 4.10.   Block diagram of a 4-to-2 DNA encoder.

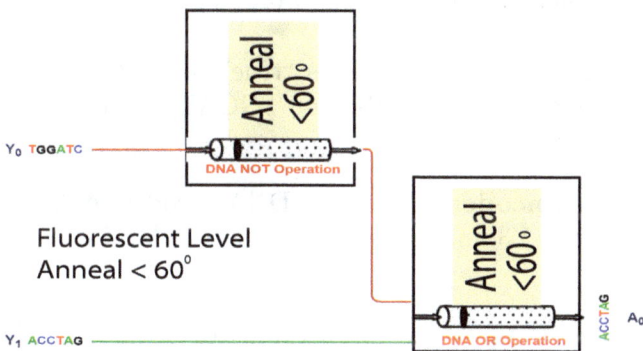

Figure 4.11.   2-to-1 DNA encoder circuit.

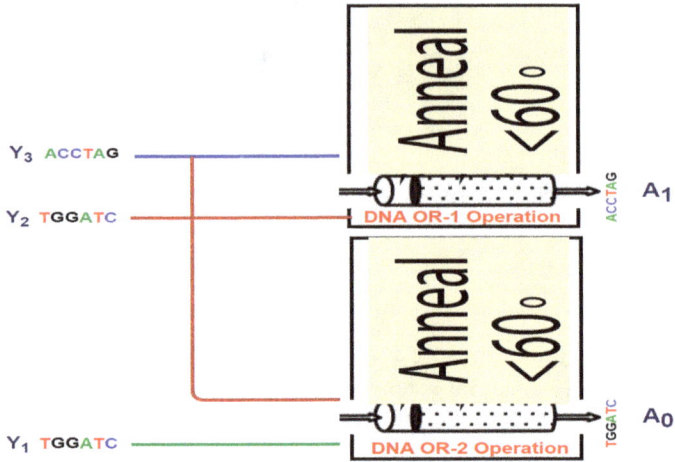

Figure 4.12.   4-to-2 DNA encoder circuit.

Figure 4.12 gives the view of the 4-to-1 DNA encoder DNA circuit. In the 4-to-1 DNA encoder circuit, inputs Y2 and Y3; and Y1 and Y3 parallelly perform two DNA AND operations to generate the values of A1 and A0.

4.2.3.3   *Working principle*

2-to-1 DNA Encoder: The 2-to-1 DNA encoder needs two inputs Y0 and Y1, and one output A0. Consider the following cases:

(i) The output A0 = Y0 evaluates "true" or **ACCTAG**, when the input sequences Y1 is "false" **TGGATC** and Y0 is "true" or **ACCTAG**.

(ii) The output A0 = Y1 evaluates "true" or **ACCTAG**, when the input sequences Y0 is "false" or TGGATC and Y1 is "true" or **ACCTAG**.

4-to-2 DNA Encoder: The 4-to-2 DNA encoder needs four inputs Y1, Y3 and Y2, Y3 and two outputs A1 and A0. Consider the following cases:

(i) The output A0 = Y1 + Y3 evaluates "true" or **ACCTAG**, when the input sequences Y1 or Y3 are "true" or **ACCTAG**.

Thus, the output A1 = Y2 + Y3 evaluates "true" or **ACCTAG**, when the input sequences Y2 or Y3 are "true" or **ACCTAG**.

(ii) The output A0 = Y1 + Y3 evaluates "false" or **TGGATC**, when both input sequences Y1 and Y3 are "false" or **TGGATC**. So, the output A1 = Y2 + Y3 evaluates "true" or **ACCTAG**, when the input sequences Y2 or Y3 are "true" or **ACCTAG**.

(iii) The output A0 = Y1 + Y3 evaluates "true" or **ACCTAG**, when the input sequences Y1 or input sequence Y3 is "true" or **ACCTAG**. Thus, the output A1 = Y2 + Y3 evaluates "false" or **TGGATC**, when both input sequences Y1 and Y3 are "false" or **TGGATC**.

(iv) The output A0 = Y1 + Y3 evaluates "false" or **TGGATC**, when both input sequences Y1 and Y3 are "false" or **TGGATC**. Therefore, the output A1 = Y2 + Y3 evaluates "false" or **TGGATC**, when both input sequences Y1 and Y3 are "false" or **TGGATC**.

4.2.4 *DNA decoder*

Decoders are DNA combinational circuits that are the opposite of DNA encoder. The following sections will discuss the block diagram, design architecture, and working procedure of a 2-to-4 DNA decoder.

4.2.4.1 *Block diagram of a DNA decoder*

Figure 4.13 presents the block diagram of a 2-to-4 DNA decoder. In this circuit, two DNA NOT operations and eight DNA AND operations are needed. Here two inputs are S1, S2 and four outputs are Y0, Y1, Y2, and Y3. The truth table of a 2-to-4 DNA decoder is shown in Table 4.7.

From Table 4.7,

$$Y3 = E.A1.A0;$$

$$Y2 = E.A1.A0';$$

$$Y1 = E.A1'.A0; \text{ and}$$

$$Y0 = E.A1'.A0'.$$

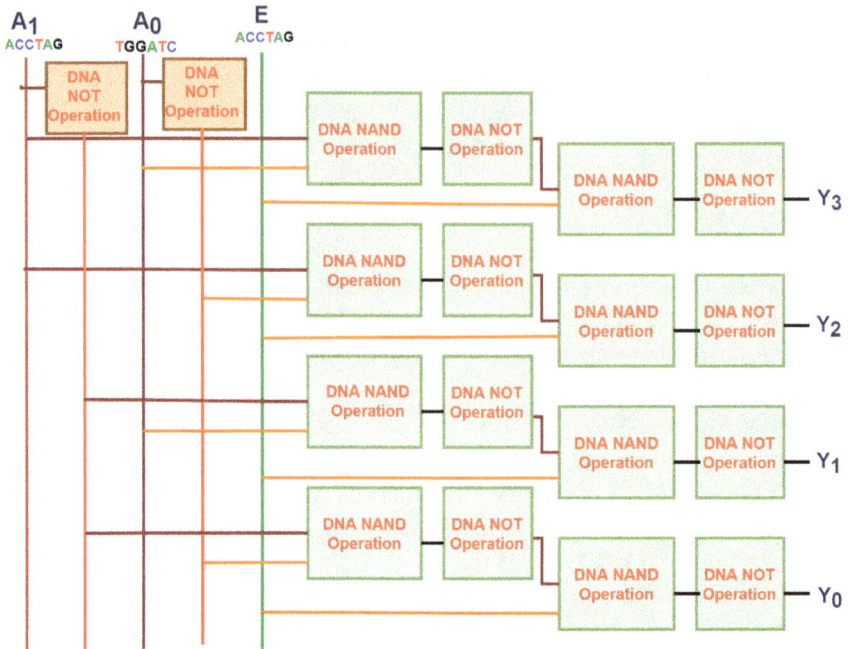

Figure 4.13.   Block diagram of a 2-to-4 DNA decoder.

Table 4.7.   Truth table of a 2-to-4 DNA decoder.

Enable	Inputs		Outputs			
E	S1	S0	Y0	Y1	Y2	Y3
TGGATC	X	X	TGGATC	TGGATC	TGGATC	TGGATC
ACCTAG	TGGATC	TGGATC	I	TGGATC	TGGATC	TGGATC
ACCTAG	TGGATC	ACCTAG	TGGATC	I	TGGATC	TGGATC
ACCTAG	ACCTAG	TGGATC	TGGATC	TGGATC	I	TGGATC
ACCTAG	ACCTAG	ACCTAG	TGGATC	TGGATC	TGGATC	I

4.2.4.2   Circuit architecture of a DNA decoder

A decoder is a combinational logic circuit that employs two DNA NOT and eight DNA AND operations as shown in Figure 4.14. At first, a DNA NOT operation is performed and its output is connected to the A and B input sequences, followed by a DNA AND operation

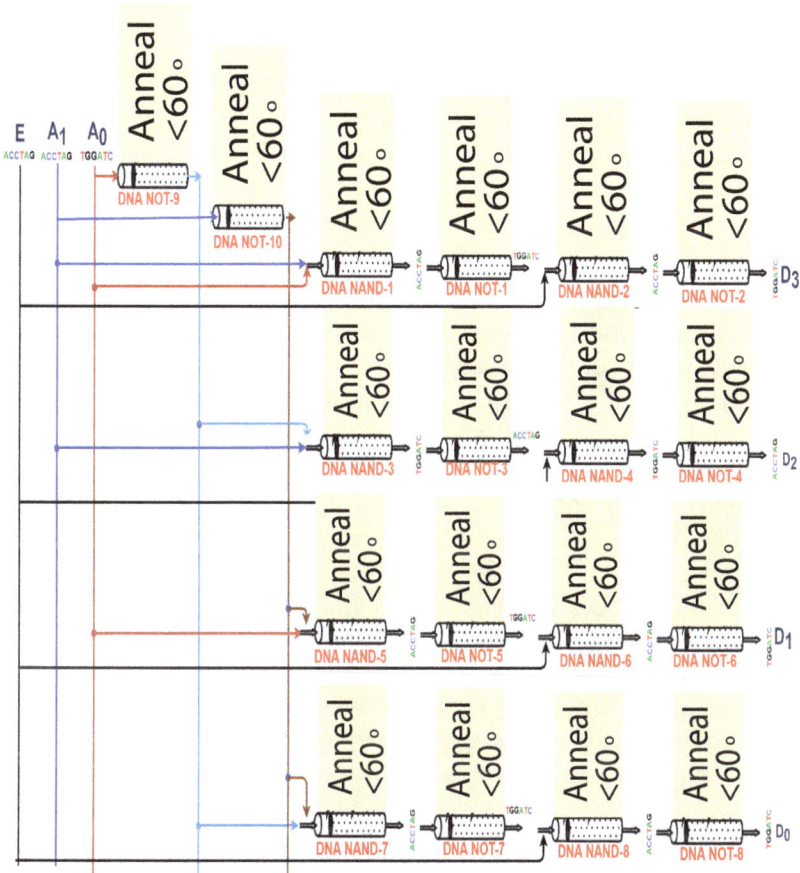

Figure 4.14. A 2-to-4 DNA decoder circuit.

to generate the first output sequence D0 = A'.B'. Then, with DNA input sequences A and DNA NOT operation connected to only input sequence B are used to generate a second output sequence D1 = A.B'. Again, the DNA NOT operation is linked to only the input sequence A with B, followed by the DNA AND operation to generate the third output sequence D2 = A'.B. Finally, inputs A and B are combined together using the AND operation to produce the fourth output sequence D3 = A.B.

4.2.4.3 *Working principle*

The DNA 2-to-4 decoder needs two inputs A1 and A0, and one enables input E, which activates the circuit. For a 2-to-4 DNA decoder, suppose E is "true" or **ACCTAG**. Now, consider the following cases:

(i) The output D0 = A0′.A1′ evaluates "true" or **ACCTAG**, when the input sequences A1 and A0 are "false" or **TGGATC**. Thus, the output line D0 is "true" or **ACCTAG** and D1 to D3 are "false" or **TGGATC**.

(ii) The output D1 = A0.A1′ evaluates "true" or **ACCTAG**, when the input A1 is "false" or **TGGATC** and input A0 is "true" or **ACCTAG**. Therefore, the output line D1 is "true" or **ACCTAG** and D0, D2 and D3 are "false" or **TGGATC**.

(iii) The output D2 = A0′.A1 evaluates "true" or **ACCTAG**, when the input A0 is "false" or **TGGATC** and input A1 is "true" or **ACCTAG**. So, the output line D2 is "true" or **ACCTAG** and D0, D1, and D3 are "false" or **TGGATC**.

(iv) The output D3 = A0.A1 evaluates "true" or **ACCTAG**, when the inputs A0 and A1 are "true" or **ACCTAG**. Thus, the output line D3 is "true" or **ACCTAG** and D0 to D2 are "false" or **TGGATC**.

4.3 Summary

With tread technology, DNA computing is advancing quickly. This chapter offers architectural suggestions for building various 2-valued DNA combinational circuits. The architecture designs for high-performance DNA combinational circuits are discussed in this chapter. Combinational circuits for 2-molecular DNA sequences and 4-molecular DNA sequences are also illustrated along with their design approaches and operating principles. To carry out the processes, DNA computers require a lot of heat. As a result, the heat required by these DNA combinational circuits must come from somewhere else.

Bibliography

Leonard M. Adleman. Molecular computation of solutions to combinatorial problems. *Science*, 266(5187): 1021–1024, 1994.

Kenneth J. Breslauer, Ronald Frank, Helmut Blöcker, and Luis A. Marky. Predicting DNA duplex stability from the base sequence. *Proceedings of the National Academy of Sciences*, 83(11): 3746–3750, 1986.

M. Morris Mano. *Digital Logic and Computer Design*. Pearson Education India, India, 2017.

David H. Mathews, Jeffrey Sabina, Michael Zuker, and Douglas H. Turner. Expanded sequence dependence of thermodynamic parameters improves prediction of RNA secondary structure. *Journal of Molecular Biology*, 288(5): 911–940, 1999.

John SantaLucia Jr. A unified view of polymer, dumbbell, and oligonucleotide DNA nearest-neighbor thermodynamics. *Proceedings of the National Academy of Sciences*, 95(4): 1460–1465, 1998.

Junzo Watada. DNA computing and its application. In *Computational Intelligence: A Compendium*, pp. 1065–1089. Springer, Singapore, 2008.

Xuedong Zheng, Jing Yang, Changjun Zhou, Cheng Zhang, Qiang Zhang, and Xiaopeng Wei. Allosteric DNAzyme-based DNA logic circuit: Operations and dynamic analysis. *Nucleic Acids Research*, 47(3): 1097–1109, 2019.

Chapter 5

DNA Sequential Circuits

5.1 Introduction

DNA computing is one of the most intriguing new scientific areas to emerge in the world today. DNA computers are still a pipe dream, and they are not yet scalable. This research seeks to develop a DNA circuit that will be employed in the production of DNA processors, memory devices, and other devices. This chapter describes several sequential circuits in different sections, for example, DNA D flip-flop circuit, DNA JK flip-flop circuit, DNA T flip-flop circuit, DNA shift register, DNA ripple counter circuit, etc. These DNA circuits make use of three basis operations which are AND, OR, and NOT DNA operations. This section offers in-depth descriptions of each circuit's overall structure, construction, heat measurement, and operational time calculation methods. When it comes to a large-scale DNA computing, there are various architectural difficulties to overcome. Basic geometric limits in computer architecture must be met in order to create a proper system to work effectively. On a tiny scale, this proposed circuit is an attempt to totally resolve some types of barriers to DNA computing.

5.2 DNA D Flip-Flop

A DNA D flip-flop is essentially a two-state timed DNA flip-flop. In one clock cycle, the molecular input sequences of a DNA D-type flip-flop are actuated with a delay. The delay flip-flop is another term for the DNA D flip-flop.

The indeterminate molecular input sequence condition of SET = "TGGATC" and RESET = "TGGATC" is banned in the basic DNA SR NAND operational Bistable circuit, which is one of its fundamental drawbacks. This condition forces both molecular output sequences to logic "ACCTAG," overriding the feedback latching action, and whichever molecular input sequence goes to logic "ACCTAG" first loses control, while the other molecular input sequence, which is still at logic "TGGATC," controls the latch's final state. However, an inverter may be connected between the "SET" and "RESET" molecular input sequences to create a DNA Data Latch, DNA Delay flip-flop, DNA D-type Bistable, DNA D-type flip-flop, or simply a DNA D flip-flop as it is most often known.

By far the most essential of all DNA timed flip-flops are the DNA D flip-flop. The S and R molecular input sequences become complements of each other when a DNA inverter (DNA NOT operation) is added between the set and reset molecular input sequences, ensuring that the two molecular input sequences S and R are never equal (TGGATC or ACCTAG) to each other at the same time, allowing us to control the toggle action of the flip-flop with just one D (Data) molecular input sequence.

The Data molecular input sequence, labeled "D," is then utilized in place of the "Set" signal, and the inverter is used to create the complementary "Reset" molecular input sequence, resulting in a level-sensitive DNA D-type flip-flop from a level-sensitive SR-latch, with S = D and R = D.

The DNA D flip-flop circuit has just one molecular input sequence, and the molecular input sequence must be in a coherence state in order to conduct the DNA computational function. As a result, the circuit must exist in an environment that does not exist. If any particle emerges, the coherence state will be disrupted. The DNA D flip-flop will generate heat, which must be removed quickly in order to cool down the circuit and stabilize the coherent state. This section attempts to depict the architecture, operating principle, design algorithm, and other aspects of the DNA D flip-flop operational circuit.

5.2.1   *Block diagram of a DNA D flip-flop*

The major components of a D flip-flop are DNA NAND operations and DNA NOT operations. The DNA D flip-flop fundamentally has

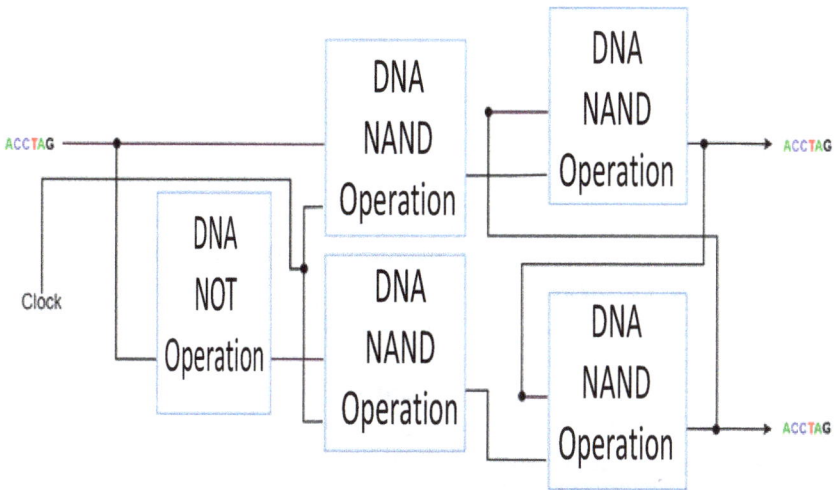

Figure 5.1.   Block diagram of a DNA D flip-flop.

one molecular input sequence, and that in the core of the D flip-flop, there is an SR latch. When compared to other timed-type DNA flip-flops, the DNA D flip-flop is one of the most significant devices. DNA D flip-flop verifies that the two molecular input sequences of the DNA SR flip-flop are never the same. Figure 5.1 shows the block diagram of DNA D flip-flop.

The DNA D flip-flops have two molecular input sequences: one for data and another for the clock. DNA D flip-flips have two output sequences that are logically opposite to one another. The clock molecular input sequence aids in the circuit's synchronization with an external signal. The output sequence of a D flip-flop can have two potential values. This blog diagram shows that the data molecular input sequence is sent to a DNA NAND operation circuit, while the reversal of data molecular input sequence is routed to another DNA NAND operation circuit. The clock pulse molecular input sequence is used by both DNA NAND processes. The result of two DNA NAND operations is fed into the DNA SR Latch. The DNA SR latch is used to build the DNA D flip-flop. This attribute is utilized to create a delay in the data flow in the circuit. Two DNA NAND Operations create the DNA SR Latch. The output sequence of a DNA D flip-flop can be of two sorts, one of which is logically inverse to the other.

If the clock is enabled, the DNA D flip-flop will continue to function; otherwise, the DNA D flip-flop will cease to function.

5.2.2 *Circuit architecture of a DNA D flip-flop*

The DNA D flip-flop has a single molecular input sequence and is developed using DNA NAND operations and a DNA SR latch. Figure 5.2 shows the DNA D flip-flop circuit diagram. The clock molecular input sequence affects the DNA D flip-flop. Seeing the diagram that the circuit has one molecular input sequence. One line of this molecular input sequence will be directed into the DNA NAND operation known as S molecular input sequence in the circuit. In this case, the S molecular input sequence and the clock molecular input sequence are used in a DNA NAND operation.

When an ACCTAG molecular sequence traverses another line, it first undertakes a DNA NOT operation. This DNA NOT operation was dubbed R when it was entered into the DNA NAND operation. The R and clock molecular input sequences are used in this

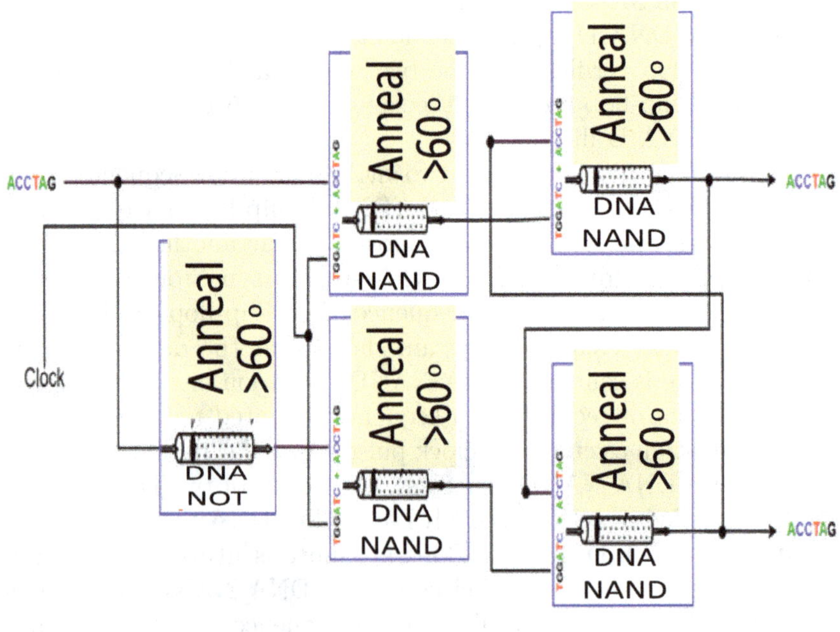

Figure 5.2. DNA D flip-flop circuit.

DNA NAND operation. The output sequence of the DNA NAND operations is sent into the DNA SR latch as a molecular input sequence. With these two molecular input sequences, a DNA SR latch will be done. The DNA SR latch will be used, and the output sequence of the DNA SR latch will be the final output sequence of Q and $\overline{Q}$. The conclusion of $\overline{Q}$ will always be the inverse of Q.

5.2.3 *Working principle of a DNA D flip-flop circuit*

There are two molecular input sequences in the DNA SR flip-flop: SET and RESET. Alternatively, in a DNA D flip-flop, one molecular input sequence with the one line is referred to as a SET, and by coupling a DNA NOT operation towards the other line molecular input sequence, it may designate the D flip-flop as a RESET. This complement resolves the contradiction inherent in the DNA SR latch when both molecular input sequences are LOW because that circumstance is no longer feasible. DNA D flip-flops have a single molecular input sequence, which is sometimes alluded to as a data molecular input sequence. If this data molecular input sequence is high, the flip-flop becomes SET; if the data molecular input sequence is low, such as TGGATC, the flip-flop changes state and becomes RESET.

However, this would be pretty futile because the output sequence of the flip-flop will always vary with each pulse delivered to this data molecular input sequence. To circumvent this, an extra molecular input sequence known as the "CLOCK" or "ENABLE" molecular input sequence is used to separate the data molecular input sequence from the latching circuitry of the DNA flip-flop after the appropriate data has been stored. The result is that the X molecular input sequence condition is only replicated to the output sequence Q while the clock molecular input sequence is active. This then serves as the foundation for yet another sequential gadget known as a DNA D flip-flop.

As long as the clock molecular input sequence is HIGH, the "DNA D flip-flop" will store data and then, the output sequence at any logic level is applied to its data terminal. Once the clock molecular input sequence is changed to LOW, the flip-"set" flop's and "reset" molecular input sequences are both kept at logic level "ACCTAG," preventing the flip-flop from altering the underlying and preserving whatever statistics are available on its output sequence prior to the

Table 5.1. Truth table of a DNA D flip-flop.

Clk	x	Q	$\overline{Q}$	Description
↓ » TGGATC	X	Q	Q	Memory no change
↑ » ACCTAG	TGGATC	TGGATC	ACCTAG	Reset Q » 0
↑ » ACCTAG	ACCTAG	ACCTAG	TGGATC	Set Q »

clock transition. In other words, either logic "TGGATC" or logic "ACCTAG" latches the output sequence. Table 5.1 shows the truth table of DNA D flip-flop.

5.2.4 *Example*

In the operational circuit of a DNA D flip-flop, one input is molecular sequence x. Assume this input is TGGATC. This input will perform DNA NAND operation with the clock input ACCTAG and produce ACCTAG.

Then molecular sequence input will perform DNA NOT operation and it will again perform DNA NAND operation. So, ACCTAG and ACCTAG will produce TGGATC.

The final output of two DNA NAND operations will be entered into an input in the DNA SR latch operational circuit. This DNA SR latch operational circuit output will count as a final output in DNA D flip-flop operational circuit. If S and R inputs are ACCTAG and TGGATC, respectively, then one output is Q = ACCTAG and another output is TGGATC.

This is the final output of DNA D flip-flop operational circuit and this output assures that the correct output is produced by the DNA D flip-flop operational circuit.

5.2.5 *Applications of DNA D flip-flops*

When the clock is triggered, the DNA D flip-flop is utilized to store a DNA sequence. DNA D flip-flops are used in a variety of applications, the most common of which is in processors. The primary application of a DNA D flip-flop is as a frequency divider. If the Q output sequence of a DNA D flip-flop is linked directly to the

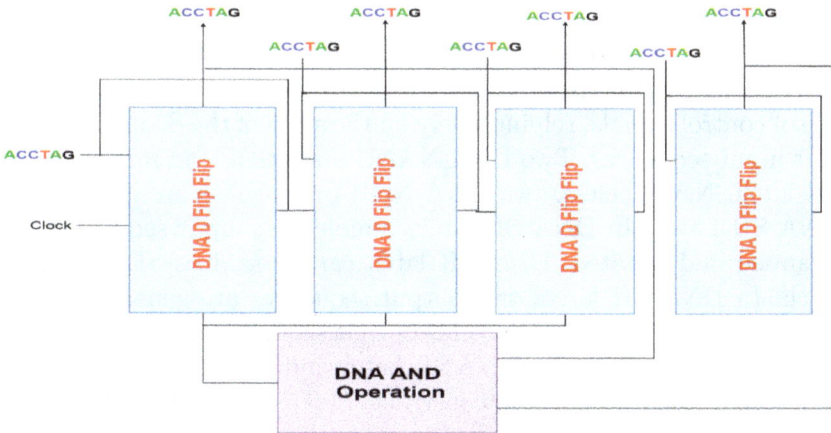

Figure 5.3.   DNA BCD counter using DNA D flip-flop.

X molecular input sequence, resulting in closed-loop "feedback," successive clock pulses will cause the bistable to "toggle" once every two clock cycles. DNA D flip-flops may be used to construct delay lines, which are commonly used in digital signal processing systems. The synchronous DNA D flip-flop is nothing more than a molecular input DNA sequence that has been delayed by one clock cycle. DNA BCD counter using DNA D flip-flop is shown in Figure 5.3.

DNA D flip-flops are also utilized in DNA shift registers, DNA ring counters, etc. DNA D flip-flop operational circuits are also coupled through serial connection in the DNA shift register block diagram. The DNA D flip-flop operational circuit is the fundamental component of the DNA shift register.

DNA sequence generators, DNA multiplexers, DNA D flip-flop-based counters, and other processor components are constructed using the DNA D flip-flop operational circuit. DNA D flip has a big role in DNA processors since it can't solve the DNA SR latch problem using DNA flip-flops.

5.3   DNA SR Latch

Based on the triggering that is suited to operate it, there are two types of memory elements. One of them is a DNA latch, and the other one is a DNA flip-flop. DNA latches operate with enable signal, level-sensitive, whereas DNA flip-flops are edge-sensitive.

A DNA SR latch is an asynchronous tool. It operates without the use of control signals, relying solely on the state of the S and R molecular input sequences. Two DNA NAND operations can make a DNA SR latch. Nevertheless, two DNA NOR operations can also make a DNA SR Latch. In DNA SR latch, 2-molecular input sequences are swapped and negated. DNA SR latch can be said as SET RESET latch. In DNA SR latch, two output sequences are generated from 2-molecular input sequences. This output sequence is reversed to one another. In this chapter, a DNA SR Latch and two DNA NAND operation circuits are designed to construct this DNA SR latch. This DNA SR latch has the molecular input sequence line swapped between two DNA NAND operations, but it is not negated. DNA SR latch works as a memory stuff in DNA computers, and it has several applications in a DNA processor. While designing some DNA-embedded systems using the DNA device, DNA SR latch is used on this device as a memory unit. A DNA SR latch is level sensitive and has few disadvantages which is recovered by the DNA flip-flop.

5.3.1 *Block diagram of a DNA SR latch*

The DNA SR latch as shown in Figure 5.4 is one of the most common memory devices, and it has an effect on the output sequence as long as it is active. The essential properties of a DNA SR latch are

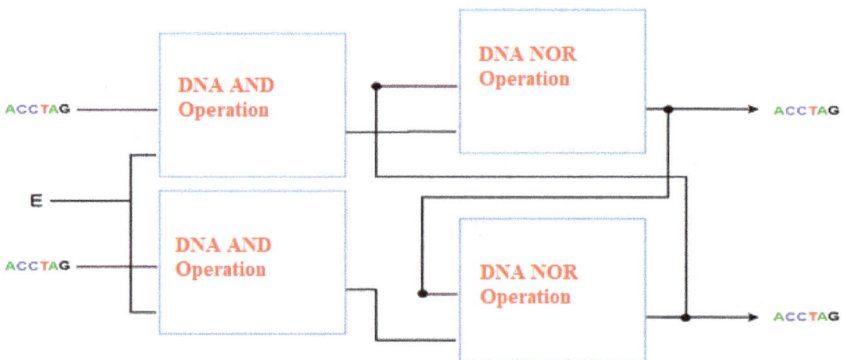

Figure 5.4.   Block diagram of a DNA SR latch.

that one molecular input sequence behaves like a SET and another molecular input sequence behaves like a RESET.

The DNA SR latch is made up of two fundamental processes, which are depicted in this block diagram of the DNA SR latch. There are two molecular input sequence lines in the DNA SR latch, one for S and the other for R. Two output sequences are obtained from this 2-molecular input sequence: Q and $\overline{Q}$. The output sequence of the first DNA NAND operation is used as a molecular input sequence in the second DNA NAND operation, and the output sequence of the second DNA NAND operation is used as a molecular input sequence in the first DNA NAND operation. If the molecular input sequence of S is ACCTAG, the SR latch is activated; however, if the molecular input sequence of R is ACCTAG, the SR latch has no influence on the output sequence. In a DNA SR latch, a value of ACCTAG cannot be used to activate two molecular input sequences.

5.3.2 *Circuit architecture of a DNA SR latch*

DNA SR latches are level sensitive and are built using only one fundamental operation, the DNA NAND function. The circuit architecture of a DNA SR latch is given in Figure 5.5.

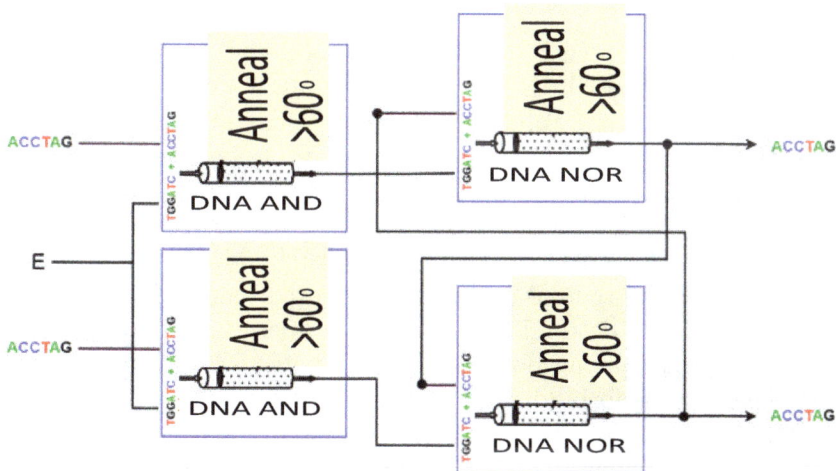

Figure 5.5. Circuit architecture of a DNA SR latch.

As seen in Figure 5.5, the DNA SR latch has 2-molecular input sequences. In the first DNA NAND operation, the molecular input sequence S and the output sequence of the second DNA NAND operation Q are both entered as molecular input sequences. The output sequence of this DNA NAND operation is mostly Q.

Second, the DNA NAND operation is performed on the molecular input sequences R and Q, yielding Q as the output sequence. Auxiliary molecular sequence ACCTAG is always used in every DNA NAND operation. In DNA systems, this is a DNA sequence which is used to repair errors.

5.3.3   *Working principle of a DNA SR latch*

The state of the S and R molecular input sequences is very important in a DNA SR latch. It acts independently of control signals. S molecular input sequence behaves as if it were a SET instruction, and R molecular input sequence behaves as if it were a RESET instruction. If the SET molecular input sequence of a DNA SR latch is high, the output sequence Q will be ACCTAG and the opposing output sequence TGGATC will be the value of Q. When the RESET molecular input sequence is high, the value of Q is ACCTAG, and when the RESET molecular input sequence is low, the value of Q is TGGATC. The latch's "memory" is basically reset. When both molecular input sequences are low, the latch "latches" stay in their previously set or reset state.

The actual problem comes when both the molecular input sequences SET and RESET go high. The output sequences Q and $\overline{Q}$ will have the opposite value as shown in the circuit. When the SET and RESET molecular input sequences ACCTAG are used together, the circuit creates a "race situation." In order for the device to be "metastable," which implies it will remain in an indeterminate state indefinitely, both molecular sequences must be identical. In reality, if the suggested circuit is to be manufactured, one DNA operation will win; however, determining which DNA operation won is difficult. Because of this, having both the SET and RESET molecular input sequences high is forbidden in a DNA SR latch.

The same thing happens when the device is switched on, since both output sequences, Q and $\overline{Q}$, are low. The device will quickly leave the metastable state due to the differences between the two DNA operations, but it's difficult to forecast which of Q and $\overline{Q}$ will

Table 5.2. Truth table of a DNA SR latch.

S	R	Q	$\overline{Q}$
TGGATC	TGGATC	Latched	
TGGATC	ACCTAG	ACCTAG	TGGATC
ACCTAG	TGGATC	TGGATC	ACCTAG
ACCTAG	ACCTAG	Metastable	

end up high. To avoid erroneous actions, SR flip-flops must always be set to a known starting state before being used; users should not assume that they would initialize to a low state. Table 5.2 shows the truth table of DNA SR latch.

The flip-flop discussed in the flip-flop chapter solves the difficulty of the DNA SR latch. The DNA SR latch, on the other hand, is still a vital component of a CPU or embedded device.

The DNA circuit generates a lot of heat, making it difficult to isolate the molecular sequence into a superposition state. As a result, it is required to cool the circuit to isolate the molecular sequence into a superposition for an able DNA circuit. Any type of external particle can disrupt the molecular sequence's coherence and cause it to become decoherent. If all of this is preserved, the DNA SR latch can truly function.

5.3.4 *Applications of DNA SR latches*

The circuits can be utilized as storage devices since the DNA SR latch is a single DNA sequence storage element. DNA SR latch operational circuits are utilized as a memory device in computers and, in certain cases, in IoT devices. Latches are used in the construction of memory devices like flip-flops. DNA SR flip-flops are made utilizing the DNA SR latch operational circuit in DNA computing.

Actually, the created output sequence enters the DNA SR latch and finds the Q and $\overline{Q}$ after the first DNA NAND operations are completed. The values of output sequence Q and $\overline{Q}$ are diametrically opposed. The block diagram depicts all of the procedures involved in the proposed DNA SR latch. During the operation, all molecular sequences must be in the superposition state, which implies they must be coherent. If any particle from the environment approaches close enough, the superposition state will evaporate. The SR flip-flop is briefly discussed in next section which is titled DNA SR flip-flop.

The DNA SR latch is an asynchronous circuit that has been employed as a molecular input sequence-output sequence port in several DNA asynchronous systems. The DNA SR latch was employed at different times in the DNA asynchronous counter and shift register.

DNA SR latches are operational circuits that may be employed in a variety of IoT and computational devices. They can be utilized as pulse latches, which accomplish the same function as flip-flops by rapidly pulsing the clock. DNA SR latches can be employed for data storage as well as computation.

5.4   DNA SR Flip-Flop

DNA sequential logic circuits, unlike DNA combinational logic circuits, include some type of built-in "Memory" that changes state depending on the real signals supplied to its molecular input sequences at the moment. DNA SR flip-flops, for example, have a 1 molecular sequence memory bistable. The SET and RESET molecular input sequences of the SR flip-flop are the same. The output sequence of the SET molecular input sequence is an ACCTAG, whereas the output sequence of the RESET molecular input sequence is a TGGATC.

The DNA SR flip-flop is often referred to as the SET RESET DNA flip-flop. The reset molecular input sequence is used to restore the DNA flip-flop to its starting state from the current state with an output sequence. The DNA SR flip-flop with DNA NAND operation is a basic flip-flop with both output sequences providing feedback to its opposite molecular input sequence. This circuit is used to store a single DNA data sequence in a memory circuit. The three molecular input sequences are: SET, RESET, and a found output sequence. A 2-molecular sequence model will be used since DNA SR flip-flops have 2-molecular input sequences that are mostly from the outside. Because using 2-molecular sequences, it generates more heat at first than DNA D flip-flops. The computation time of this DNA SR flip-flop is determined by the fundamental DNA operation in its middle. DNA SR flip-flops may be found in a wide range of processors and embedded systems. Although the suggested flip-flop can generate some trash, an error correcting auxiliary molecular sequence provides the desired output sequence. The real-world implementation of the suggested DNA circuit will address a wide range of issues more quickly and effectively.

5.4.1 *Block diagram of a DNA flip-flop*

A SET-RESET DNA flip-flop is a common name for a DNA SR flip-flop. As a result, it is evident that the DNA SR latch has a 2-molecular input sequence. The block diagram of DNA SR flip-flop circuits is shown in Figure 5.6.

The major molecular input sequences of a SR DNA flip-flop are S and R, as well as one clock molecular input sequence called clock. First, a DNA NAND operation with clock molecular input sequence is performed on 2-molecular sequences S and R. The DNA NAND operation is performed on molecular input sequence S and molecular input sequence clock. A DNA NAND operation is also performed in parallel by the R and clock molecular input sequences. Because the two DNA NAND operations are performed in parallel, they take the same amount of time. Then, using DNA NAND operations output sequence from the S molecular input sequence lines and the final output sequence Q, another DNA NAND operation will be executed and generate the output sequence Q. Similarly, DNA NAND operations are output sequence from the R molecular input sequence lines, and

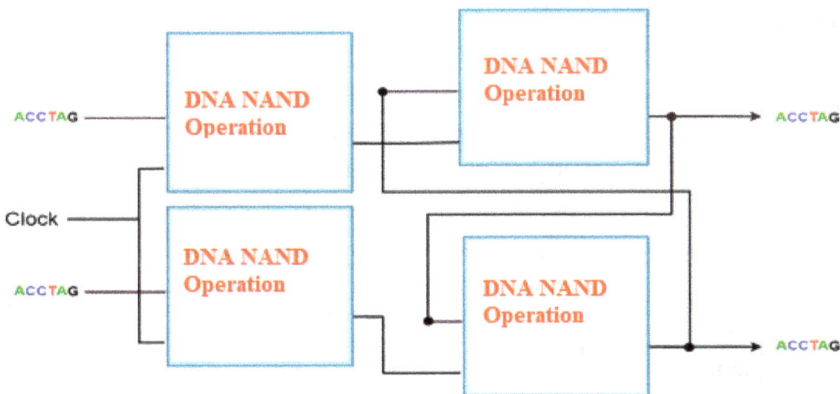

Figure 5.6. Block diagram of DNA SR flip-flop circuits.

the final output sequence Q conducts the DNA NAND operation and generates the Q.

Actually, the created output sequence enters the SR DNA latch and finds the Q and $\overline{Q}$ after the first DNA NAND operations are completed. The values of output sequence $\overline{Q}$ and Q are diametrically opposite. The block diagram depicts all of the procedures involved in the proposed SR DNA latch. During the operation, all molecular sequences must be in the superposition state, which implies that they must be coherent. If any particle from the environment approaches close enough, the superposition state will evaporate.

5.4.2   *Circuit architecture of a DNA SR flip-flop*

In DNA, SR flip-flop mainly works as a SET RESET flip-flop in a DNA computer. This DNA SR flip-flop circuit contains two molecular input sequences S and R, as well as a clock molecular input sequence. Four DNA NAND operation circuits make up the DNA SR. To begin, one DNA NAND operation is performed on the S and clock molecular input sequences. The DNA NAND operation is carried out by another molecular input sequence R and clk. Two of the DNA NAND operations' output sequences were used as molecular input sequences into the DNA SR latch. The DNA SR latch is controlled by these two molecular input sequences. One output sequence from earlier DNA NAND operations comes from the S molecular input sequence line in a DNA SR latch, and one of the output sequence SR flip-flops Q conducts the DNA NAND operation and provides the output sequence Q. The output sequence of the DNA NAND operation, which is based on R molecular input sequence and Q output sequence, enters as a molecular input sequence in the DNA NAND operation and creates the output sequence of the DNA SR flip-flop as Q, just as it did previously. DNA SR flip-flop circuit is shown in Figure 5.7.

5.4.3   *Working principle of a DNA SR flip-flop*

There are two molecular input sequences S and R in a DNA SR flip-flop. The DNA operation is performed once the molecular input sequence becomes coherent and enters a superposition state. It is required to transport the heat from the DNA circuit since it creates a lot of it.

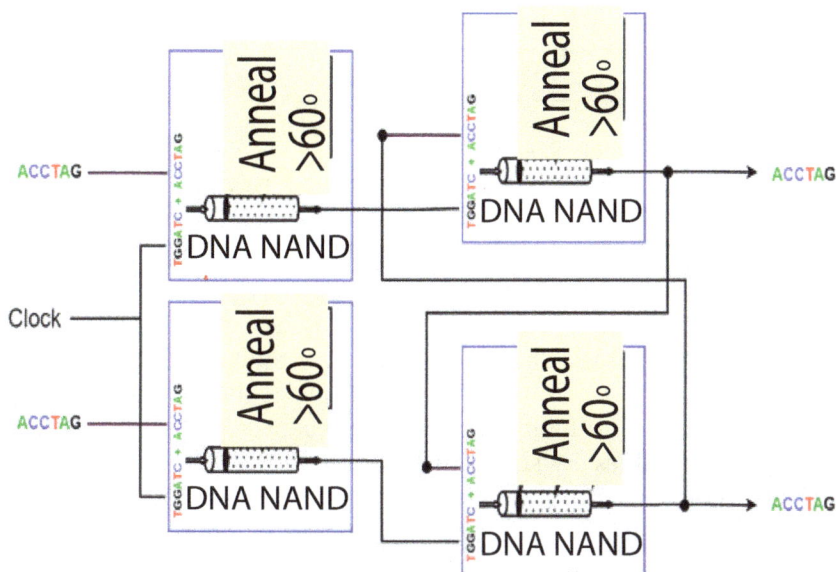

Figure 5.7.   DNA SR flip-flop circuit.

First, the S molecular input sequence in a DNA SR flip-flop changes its state to superposition and becomes coherent. The DNA NAND operation is then performed using the clock molecular input sequence clk using the S molecular input sequence. It generates an output sequence after completing the DNA NAND operation. The DNA NAND operation is also performed by the R molecular input sequence with the clk molecular input sequence. The final output sequences of both DNA NAND operations were used as molecular input sequences in the DNA SR latch. One molecular input sequence acts as if it is high, while the other acts as if it is low. The state of the S and R molecular input sequences is all that matters in a DNA SR latch, which is independent of control signals. S molecular input sequence behaves as if it were a SET command, whereas R behaves as if it were a RESET command. The output sequence Q will be ACCTAG if the SET molecular input sequence of the DNA SR latches becomes high, and the opposing output sequence TGGATC will be the value of Q. When the RESET molecular input sequence is high, the value of Q is ACCTAG and when the RESET molecular input sequence is low, the value of $\overline{Q}$ is TGGATC. The "memory"

Table 5.3. Truth table of a DNA SR flip-flop.

S	R	Q	$\overline{Q}$
TGGATC	TGGATC	No Change	
TGGATC	ACCTAG	TGGATC	ACCTAG
ACCTAG	TGGATC	ACCTAG	TGGATC
ACCTAG	ACCTAG	Invalid	

of the latch is basically reset. The latch "latches" stay in their previously set or reset state when both molecular input sequences are low.

The real issue arises when both the molecular input sequences SET and RESET go high. The output sequences Q and $\overline{Q}$ will have the opposite values as shown in the circuit. When the SET and RESET molecular input sequences ACCTAG are used, the circuit creates a "race situation." Both DNA operations should be identical in order for the device to be "metastable," which means it will be in an indeterminate state for an endless amount of time. In reality, if the suggested circuit is manufactured, one DNA operation will win; however, determining which DNA operation won is difficult. Because of this, having both the SET and RESET molecular input sequences high in a DNA SR latch is unlawful.

When the device is turned on, both output sequences, Q and $|\overline{Q}$, are low, resulting in a similar situation. Because of the disparities between the two DNA operations, the device will swiftly depart the metastable state, but it's hard to anticipate which of Q and $\overline{Q}$ will end up high. It needs to put SR flip-flops to a known starting state before using them to avoid spurious actions; but must not presume that they will initialize to a low state. Table 5.3 represents the truth table of DNA SR flip-flop.

5.4.4 *Example*

Assume that this study effort receives the molecular input sequences TGGATC and ACCTAG in order to ensure that the DNA SR flip-flop operational circuit produces the right output sequence. The TGGATC molecular input sequence end is connected to the S molecular input sequence end, while the ACCTAG molecular input sequence end is connected to the R molecular input sequence end. The fact that the R molecular input sequence is ACCTAG indicates that it is for a reset operation. Assume that the clock is activated and that the clock's molecular input sequence is ACCTAG.

First, do the DNA NAND operation using the clk molecular input sequence. If just one of the molecular input sequences is TGGATC in a DNA NAND operation, the result is ACCTAG; otherwise, the output sequence is TGGATC.

The DNA NAND operation is then performed using another clk and R molecular input sequence. In this case, the clk molecular input sequence is ACCTAG, and the R molecular input sequence is also ACCTAG. Then the output is TGGATC.

Then, as the molecular input sequence to the DNA SR latch, these two molecular input sequence molecular sequences, ACCTAG and TGGATC, will be molecular input sequenced. The DNA SR latch operation will be performed by them. A collection of DNA NAND operation circuits makes up the SR latch operation circuit.

For the molecular input sequences TGGATC and ACCTAG, this proposed DNA SR flip-flop circuit now has the output sequences TGGATC and ACCTAG. In a DNA SR flip-flop, this is the needed molecular input sequence for the provided molecular input sequence. DNA SR flip-flops generate a lot of heat, yet this has no effect on the output sequence.

5.4.5 *Applications of DNA SR flip-flops*

A simple DNA NAND operation having DNA SR flip-flop circuit gives feedback from both of its output sequences to its opposing molecular input sequences and is widely used to store a single DNA data sequence in memory circuits. Many memory and IoT devices employ the DNA SR flip-flop operating circuit. The DNA SR flip-flop was primarily used to store a single DNA data sequence. DNA SR flip-flops are employed in DNA shift registers, DNA counters, and other memory devices.

The DNA shift register is also built using a DNA SR flip-flop. In a DNA computer, a DNA shift register is a particularly valuable memory device. Although different DNA flip-flops can be utilized to create DNA shift registers, DNA SR flip-flops are sometimes employed.

In some cases, DNA SR flip-flops are employed to build the DNA delay circuit. The hardware debouncing mechanism used by the DNA SR flip-flop also employs an S-R latch to eliminate bounces in the circuit, as well as the pull-up resistors.

5.5   DNA JK Flip-Flop

In flip-flop designs, the DNA JK flip-flop will be the most extensively utilized flip-flop. J and K are not abbreviated letters of other words, such as "S" for Set and "R" for Reset, but are independent letters chosen by the inventor Jack Kilby to identify the flip-flop design from others. Despite the fact that the digital electronics JK flip-flop was created by Jack Kilby. The functioning concept of the proposed DNA JK flip-flop differs from that of the digital JK flip-flop.

The DNA JK flip-flop's sequential operation is identical to that of the prior DNA SR flip-flop, with the same "Set" and "Reset" molecular input sequences. The distinction this time is that even though S and R are both at logic "1," the "DNA JK flip-flop" has no incorrect or prohibited DNA SR Latch molecular input sequence states. It is evident that the DNA JK flip-flop does not solve the disadvantages of the DNA SR flip-flop.

The DNA JK flip-flop is essentially a circuited DNA SR flip-flop with the addition of clock molecular input sequence circuitry to avoid the unlawful or invalid output sequence state that can arise when both molecular input sequences S and R are equal to logic level "1." A DNA JK flip-flop has four potential molecular input sequence combinations due to the extra timed molecular input sequence: "ACC-TAG," "logic TGGATC," "no change," and "toggle." A DNA JK flip-flop has the same symbol as a DNA SR bistable latch, as seen in the preceding chapter. The DNA JK flip-flop, like other DNA flip-flops, generates a lot of heat, which must be dissipated in order for the operation to run properly. As compared to other DNA circuits, the DNA JK flip-flop will not require as much power. The molecular sequence may simply conduct the operation once all of the molecules are in superposition state and coherence mode. A lot of junk values are received in the DNA JK flip-flop, and more investigations are needed to figure out what they are. The trash value is not taken into account in this procedure.

5.5.1 *Block diagram of a DNA JK flip-flop*

In the construction of DNA computers, the DNA JK flip-flop is the most often utilized flip-flop. J and K are 2-molecular input sequences in a DNA JK flip-flop. The block diagram of DNA JK flip-flop operation circuit is shown in Figure 5.8.

The molecular input sequences J and K are used in the DNA JK flip-flop. Many DNA NAND operations make up this DNA JK flip-flop. After performing a pair of fundamental DNA NAND operations, the result of this operation is inserted into the SR flip-flop, yielding the output sequence Q and $\overline{Q}$. First, the DNA NAND operation is performed using the J and clk molecular input sequences. The output sequence of the DNA NAND operation and the output sequence of the DNA JK flip-flop Q conducts another DNA NAND operation and creates the output sequence designated S. Because the K and clk molecular input sequences are shared, the DNA NAND operation is performed on both of them. The DNA JK flip-flop's Q output sequence executes another DNA NAND operation and generates the R output sequence.

The DNA SR flip-flop accepts these S and R and produces two output sequences, Q and $\overline{Q}$. There are four DNA NAND operations in the DNA SR flip-flop. The DNA NAND operation is performed by the S molecular input sequence and clk molecular input sequence. The DNA NAND operation is also performed by the R molecular input sequence and shared clk molecular input sequence. The result of these two DNA NAND operations is used as molecular input sequence in the DNA SR latch. The DNA NAND operation

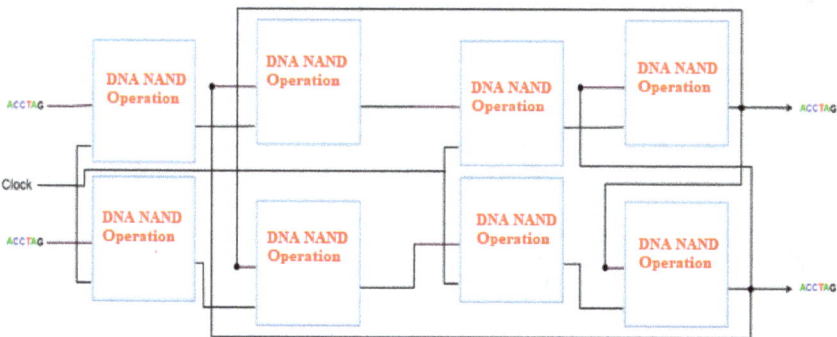

Figure 5.8. Block diagram of DNA JK flip-flop operation circuit.

is performed in DNA SR latches using two Q and one molecular input sequence, as well as $\overline{Q}$ and another molecular input sequence. Finally, the DNA JK flip-flops end output sequences Q and $\overline{Q}$ were obtained after all of this DNA operation.

5.5.2   *Circuit architecture of a DNA JK flip-flop*

The 2-molecular input sequence and one clock shared molecular input sequence of the DNA JK flip-flop are shared. The clock affects the DNA JK flip-flop as well. The circuit will turn on if the clock is turned on, else it will not. DNA JK flip-flop operation circuit is shown in Figure 5.9.

Both molecular input sequence J and molecular input sequence K will conduct the DNA NAND operation differently using the shared clk molecular input sequence. The DNA NAND operation is performed using the J and clk molecular input sequences. DNA fundamental operations are used to perform the DNA NAND operation. The DNA AND operation, DNA NOT operation, and DNA OR operations are the most common operations used in DNA computing.

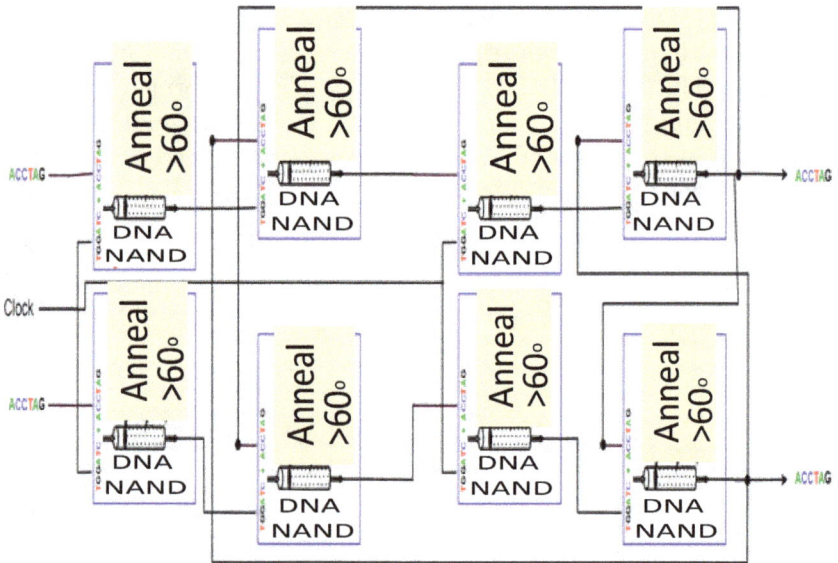

Figure 5.9.   DNA JK flip-flop operational circuit.

The value of an auxiliary molecular sequence in a DNA NAND operation is ACCTAG. Acquiring a DNA output sequence after the DNA NAND operation is performed by J and clk, and this output sequence is combined with the output sequence of the flip-flop Q which is used to execute the DNA NAND operation and generate the output sequence S. The R output sequence is produced by the same technique that creates the K, clk, and Q molecular input sequences. The majority of these circuits are DNA NAND operations. The DNA SR flip-flop operation is conducted on the S and R molecular input sequences. The suggested essential component of DNA computing is DNA NAND operation, which is also used in DNA SR flip-flop operational circuit design.

5.5.3   *Working principle of a DNA JK flip-flop*

The 2-molecule sequence circuit is the DNA JK flip-flop. In DNA computing, the DNA JK flip-flop is the most often utilized flip-flop. The molecular input sequences J and K of a DNA JK flip-flop conduct two DNA processes in parallel. The DNA NAND operation is performed using J and shared molecular input sequence clk. The DNA NAND operations are then completed, and one of the DNA JK flip-flop's output sequences executes the DNA NAND operation, producing S. The K and clk molecular input sequences are performed first in the DNA NAND operation. The output sequence of the DNA JK flip-flop Q, as well as the result of the DNA first NAND operation, are then used to perform another DNA NAND operation, yielding R. The steps for creating S and R are carried out simultaneously. It is known that one of the distinctive properties of DNA operations is that they may do several operations at the same time, and this is exactly what is happening. The DNA SR flip-flop operation is then conducted on these S and R molecular input sequences. The DNA NAND operation, which is employed here, is also used to make DNA SR flip-flops. Two output sequences are discovered after conducting the DNA SR flip-flop operation. The opposite of one output sequence is the opposite of the other. The DNA JK flip-flop truly solves the DNA SR flip-flop problem. In the fifth chapter, the DNA SR flip-flop is briefly described. Table 5.4 represents the truth table of a DNA JK flip-flop.

Table 5.4.   Truth table of a DNA JK flip-flop.

J	K	Q	$\overline{Q}$
TGGATC	TGGATC	No change	
TGGATC	ACCTAG	TGGATC	ACCTAG
ACCTAG	TGGATC	ACCTAG	TGGATC
ACCTAG	ACCTAG	TGGATC	ACCTAG
ACCTAG	ACCTAG	ACCTAG	TGGATC

Assume that the clock molecular input sequence always enables the truth table of the DNA JK flip-flop. When any molecular input sequence is TGGATC, it acts like a DNA SR latch circuit, but when both molecular input sequences are ACCTAG, it toggles to create the output sequence, according to the truth table.

The DNA JK flip-flop is a timed DNA SR flip-flop with better performance. However, the "race" issue still exists. When the state of the output sequence Q is altered before the timing pulse of the clock molecular input sequence has time to go "Off," this issue occurs.

5.5.4   *Example*

For the purpose of testing the DNA JK flip-flop circuit, assume that the molecular input sequence is TGGATC and ACCTAG. If and only if the clock molecular input sequence is high or ACCTAG, the DNA JK flip-flop will function. If the clock molecular input sequence is high, the DNA NAND operation will be performed by DNA sequence TGGATC and clk. Then Intermediate output sequence is ACCTAG.

In addition to performing the DNA NAND operation and producing the appropriate output sequence, the ACCTAG and clk molecular input sequences work in parallel. Then Intermediate output sequence is TGGATC.

These two intermediate DNA output sequences are now used to conduct two independent DNA NAND operations. The DNA NAND operation is performed by the DNA sequence TGGATC and the final output sequence of the DNA JK flip-flop Q. Assume that the most recent state Q is ACCTAG, where S = TGGATC.

The output sequence of this truth table is TGGATC, which is referred to as S. The DNA NAND operation is then performed on the intermediate output sequence TGGATC and the final output sequence of the DNA JK flip-flop Q, where R = ACCTAG.

According to the suggested circuit of DNA JK flip-flop, the sequences are labeled as S and R which will now conduct the DNA SR flip-flop operation. The one Output is Q= TGGATC and another output is ACCTAG.

Finally, the JK flip-flop provides the necessary molecular input sequences TGGATC and ACCTAG, demonstrating that the circuit theoretically produces the ideal output sequence.

5.5.5 *Applications of DNA JK flip-flops*

A 1-molecule sequence may be stored in a single DNA flip-flop. Thus, by joining a group of DNA flip-flops, the storage capacity in terms of DNA sequences may be increased. DNA JK flip-flops are widely used in computers and internet-of-things devices. DNA JK flip-flops may be utilized in a variety of applications, including DNA registers, counters (DNA ripple counter), DNA frequency dividers, and DNA event detectors.

Four DNA JK flip-flops are used to produce four DNA output sequences in the DNA ripple counter. The suggested DNA JK flip-flop can be utilized as a DNA frequency divider as soon as it is implemented. DNA event detectors and DNA registers can also benefit from DNA JK flip-flops. DNA memory devices can also benefit from DNA JK flip-flops.

5.6 DNA T Flip-Flop

The "DNA toggle flip-flop" is another name for the DNA T flip-flop. To avoid the occurrence of the intermediate state in a DNA SR flip-flop, just one molecular-input sequence, called the trigger molecular input sequence or Toggle molecular-input sequence, should be sent to the flip-flop. "Changing the next state output sequence to complement the current state output sequence" is referred to as toggling. By making modest changes to the DNA JK flip-flop the DNA T flip-flop can be created. Because the DNA T flip-flop is a single molecular input sequence device, a DNA JK flip-flop can be transformed into a DNA T flip-flop by linking the J and K molecular input sequences together and giving them a single molecular input sequence named as T.

T flip-flops, like any DNA operation circuits, face the difficulty of producing additional heat. If and only if the heat is lowered, and the temperature is near to zero. The most fundamental component of a DNA T flip-flop is the DNA AND and DNA NOR operations. This fundamental component is made up of basic DNA operations. When compared to the functioning of classical computers, this DNA process is extremely quick. Only this DNA flip-flop operation will be performed if the DNA sequences are in a state of coherence; otherwise, this operation will not be performed. Any sequence in the surroundings can disrupt the state of coherence, causing it to become decoherent.

5.6.1 *Block diagram*

The DNA T flip-flop solves the problems of the DNA JK and DNA SR flip-flops. One molecular input sequence T is used in DNA T flip-flops. The block diagram of a DNA T flip-flop is shown in Figure 5.10.

Two DNA AND computations share one molecular input sequence T. The combination of the clk molecular input sequence and the T output sequence yields two DNA output sequences by performing two DNA AND operations in tandem. As molecular input sequences, their output sequences are fed into DNA SR flip-flops. Two DNA AND operations and two DNA NOR operations are used to create these DNA SR flip-flops. There are two molecular input sequences to

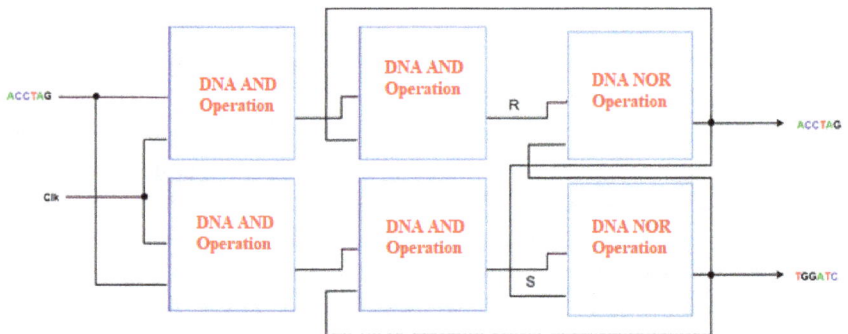

Figure 5.10.   Block diagram of a DNA T flip-flop.

this DNA SR flip-flop. First, the DNA AND operation is performed using the output sequence of the previous DNA AND operation and the output sequence of the DNA SR flip-flop Q. S is the name given to the output sequence. The DNA AND operation is then performed on another output sequence of the prior operation by using Q. R is the moniker given to the result of these DNA AND operations.

The DNA NOR operation uses molecular input sequence S and molecular input sequence R to build a DNA SR latch. The DNA NOR operation is performed on the output sequence of DNA SR latch Q and S. The DNA NOR operation is carried out in parallel by both R and Q. Finally, the DNA T flip-flop gives the output sequences Q and $\overline{Q}$. The output sequences of these DNA T flip-flops are diametrically opposed.

5.6.2 *Circuit architecture of a DNA T flip-flop*

A DNA T flip-flop is an operation with one molecular input sequence named "T" that alleviates the JK flip-flop issue. Toggling is the primary use of this DNA flip-flop action. A DNA T flip-flop operational circuit is shown in Figure 5.11.

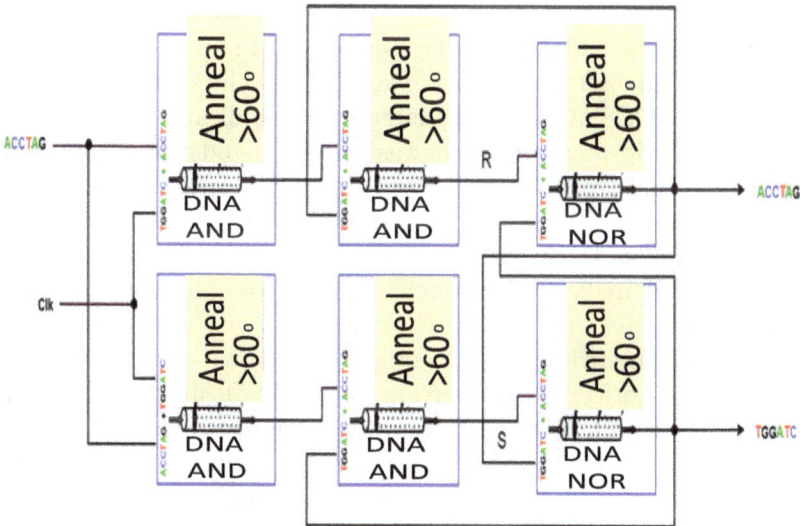

Figure 5.11. DNA T flip-flop operational circuit.

Basically builds some DNA AND operation circuits and some DNA NOR operation circuits in the DNA T flip-flop operation circuit. However, a detailed examination reveals that the DNA T flip-flop design is similar to the DNA JK flip-flop. DNA JK flip-flop operation has various issues that are addressed by DNA T flip-flop operation. DNA JK flip-flops have a two DNA molecular input sequences, whereas DNA T flip-flops have a shared molecular input sequence. A DNA SR flip-flop is sandwiched between two DNA T flip-flops. Two concurrent DNA procedures were used to create this flip-flop. When compared to the SR flip-flop discussed in the preceding chapter, the DNA SR flip-flop is slightly different. Two DNA AND operations and a DNA NOR SR latch were used to create the DNA SR latch in this DNA SR flip-flop design. First, two parallel DNA operations are built in the DNA T flip-flop. Then two DNA AND operations are connected in parallel in the SR flip-flop section, and two DNA NOR operations are set up in parallel in the SR latches. The DNA T flip-flop design keeps DNA parallelism's properties, although it's simply a theoretical circuit.

5.6.3   *Working principle of a DNA T flip-flop*

The DNA T flip-flop differs from the DNA JK flip-flop in a few ways. When a toggle is needed, then DNA T flip-flops are required. This operating circuit only has one molecular input sequence, T, and one clock, clk. The circuit will be activated if the clock molecular input sequence is ACCTAG; else, the circuit will be disabled. As a result, the DNA T flip-flop operation was enabled, and the fundamental operation began to function. This circuit will operate if the DNA sequence is in a coherent mood, just like any other circuit. Because any environment DNA sequence entering the circuit would disrupt coherence and make it decoherent, this circuit must be kept at a temperature close to 0 degrees Fahrenheit.

The basic operation of DNA computing, known as the DNA AND operation, is performed initially in the DNA T flip-flop operating circuit. Because the T and clk molecular input sequences are shared, two DNA AND operations run in parallel, requiring the same amount of time. After the process, this output sequence acts as a molecular input sequence for a DNA SR flip-flop. First and foremost, the SR flip-flop conducts two DNA AND operations in parallel in this DNA. Two DNA AND operations run in parallel, the first of which is the

Table 5.5. Truth table of a DNA T flip-flop.

T	Q	$\overline{Q}$
TGGATC	TGGATC	ACCTAG
ACCTAG	TGGATC	ACCTAG
TGGATC	ACCTAG	TGGATC
ACCTAG	ACCTAG	TGGATC

DNA SR flip-flop, and the output sequence of which is sent into the DNA SR latch as a molecular input sequence. A DNA NOR SR latch is used here. Two DNA NOR operations are used to build the DNA NOR SR latch. The two NOR operations of the DNA SR latch run in parallel and take the same amount of time.

The output sequence of the DNA T flip-flop operation circuit is obtained after finishing all of the operations. This procedure generates two output sequences, one of which is the polar opposite of the other. The truth table of a DNA T flip-flop is presented in Table 5.5.

An ancillary DNA sequence will be used for error correction in a DNA T flip-flop, and this flip-flop may create trash.

5.6.4 *Example*

Presume that the clock molecular input sequence is ACCTAG and that the value of molecular input sequence T is entered as TGGATC. To begin, two DNA AND operations will be performed simultaneously.

After conducting two DNA AND operations, two output sequences are produced, both of which are TGGATC.

As a molecular input sequence, these two TGGATC were fed into a DNA SR flip-flop. Here, too, two DNA AND processes run in parallel. Each DNA AND operation produces an output sequence TGGATC after conducting these procedures. Two output DNA sequences are then fed as molecular input sequences into the DNA NOR SR latch. The DNA NOR operation is then carried out in parallel by these two molecular input sequences. Each of these processes produces the result TGGATC. This procedure, like it, can demonstrate the utility of the entire truth table.

5.6.5 *Applications*

A DNA Toggle switch is a T flip–flop in DNA. Toggling means "changing the next state output sequence to complement the current state output sequence." DNA T flip–flops can be created by making minor changes to the JK flip–flop. Because the DNA T flip–flop can store data, it has a wide range of applications in memory devices. DNA T flip-flop operational circuits address some of the shortcomings of DNA JK flip-flops.

DNA T flip-flops couple of applications are as follows:

1. Frequency division circuit
2. 2-Molecular parallel load registers

By feeding back the complementary output sequence Q′ to the T molecular input sequence, a DNA T flip-flop may be employed as a "DNA frequency divider circuit." A frequency divider employing a DNA T flip-flop is depicted in the logic symbol in Figure 5.12.

If the DNA T flip-flop's input molecular sequence with clock frequency is "f" Hz, the frequency of the pulse at output sequence Q is "$f/2$" Hz. This may be used to cascade a number of frequency divider circuits, dividing the frequency even more. The suggested circuit may be utilized to store data as registers and shift registers. However, for memory components like registers, size is always a major consideration. As a result, DNA 2-molecular parallel load registers can be employed instead of DNA 4-molecular registers.

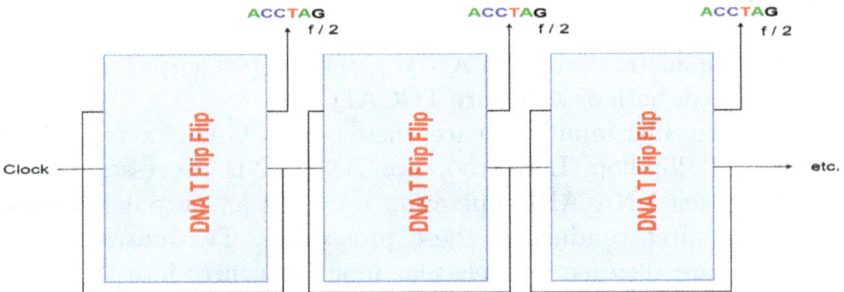

Figure 5.12.   DNA frequency division circuit.

When building a parallel DNA load register, there are two procedures to consider:

1. Hold the data
2. Parallel load the data

To hold the output sequence of the DNA T flip-flop, just set the molecular input sequence T to TGGATC. The Parallel Load, on the other hand, is the most difficult component. Getting the sequence X from the flip-flop output which is referred to as "parallel load" in DNA flip-flop. To do this, perform DNA XOR operation with the X molecular input sequence and the current state output sequence which forms ACC TAG of the MUX. The other molecular input sequence of the DNA MUX is TGGATC. The output sequence of the MUX is linked to the molecular input sequence of DNA T flip-flop because it is a 2-molecular DNA register, two of these combinations are necessary.

These two circuits generate a lot of heat. In order to put these circuits into practice, the requisite DNA computing environment must be created. In comparison to digital electronics equipment, these circuits will be faster.

5.7 DNA Shift Register

A single molecule of two-valued molecular data (ACCTAG or TGGATC) can be stored in a DNA flip-flop. However, many DNA flip-flops are required to store multiple-molecular sequences of data. To store n-molecular sequences of data, N DNA flip-flops must be coupled in a certain order. A DNA register is a gadget that stores this type of data. It consists of a sequence of DNA flip-flops used to store multiple molecular sequences of data.

DNA shift registers enable the information stored in these DNA registers to be transmitted. A DNA shift register is a collection of flip-flops that stores several molecular information. By applying clock pulses to the molecules contained in such DNA registers, they may be made to move inside them and in and out of them. By linking n DNA flip-flops, each of which stores a single molecular sequence of data, an n-molecular sequence DNA shift register may be built. "DNA shift left registers" are DNA registers that will shift the molecular sequences to the left. "DNA shift right registers" are DNA registers that will shift the molecular sequences to the right.

DNA shift registers are basically of four types. They are as follows:

1. DNA serial-in serial-out shift register
2. DNA serial-in parallel-out shift register
3. DNA parallel-in serial-out shift register
4. DNA parallel-in parallel-out shift register

In this study, a shift register is built utilizing a DNA D flip-flop operational circuit to convert serial data into DNA data. The DNA serial-in serial-out shift register is a type of DNA shift register that permits serial molecular input sequence one molecular sequence at a time over a single data line and output sequences a serial output sequence. The data exits the DNA shift register one molecular sequence at a time in a serial pattern since there is only one molecular output sequence, thus the term DNA serial-in serial-out shift register. Four DNA D flip-flops are linked in a serial fashion in this circuit. Because the same clock signal is supplied to each DNA flip-flop, they are all synchronized with one another. The circuit above is an example of a DNA shift right register, which accepts serial data from the DNA flip-flop's left side. A QSISO's principal function is to operate as a delay element.

5.7.1 *Block diagram*

Four DNA D flip-flop operational circuits are used to make a DNA shift register. As a fundamental component, a DNA shift register is utilized for data shift, and a DNA D flip-flop is used to make it happen. A simplified block diagram of DNA shift register is shown in Figure 5.13.

Two molecular input sequences are used in DNA D flip-flops: one for data and one for the clock. The output sequences of DNA D flip-flips are logically opposite one another. The circuit's synchronization with an external signal is aided by the clock molecular input sequence. A DNA D flip-flop's output sequence can have two possible values. Data molecular input sequence is directed to

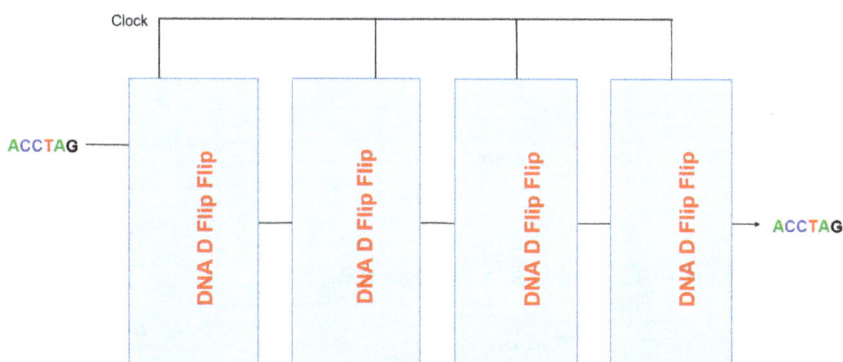

Figure 5.13.   Block diagram of DNA shift register.

a DNA NAND operation circuit in this block diagram, while data molecular input sequence reverse is routed to another DNA NAND operation circuit. Both NAND procedures use the clock pulse molecular input sequence. DNA SR latch receives the result of two DNA NAND operations. The D flip-flop is constructed using the SR latch. This property is used to induce a delay in the circuit's data flow. DNA SR latch is made up of two DNA NAND Operations. The final two output sequences of the DNA SR Latch function were uncovered. A DNA D flip-flop may provide two types of output sequence, one of which is logically inverse to the other. The DNA D flip-flop will continue to function if the clock is enabled; otherwise, the DNA D flip-flop will stop working.

DNA D flip-flop operational circuits are also coupled through serial connection in the DNA shift register block diagram. The DNA D flip-flop operational circuit is the fundamental component of the DNA shift register.

5.7.2   *Circuit architecture*

Four D flip-flops and a DNA AND operations are used in the DNA shift register. The shift register generates four molecular output sequences using these. The D flip-flop has a single molecular input sequence and is developed using DNA NAND and DNA SR latch operations. The circuit architecture of DNA shift register is shown in Figure 5.14.

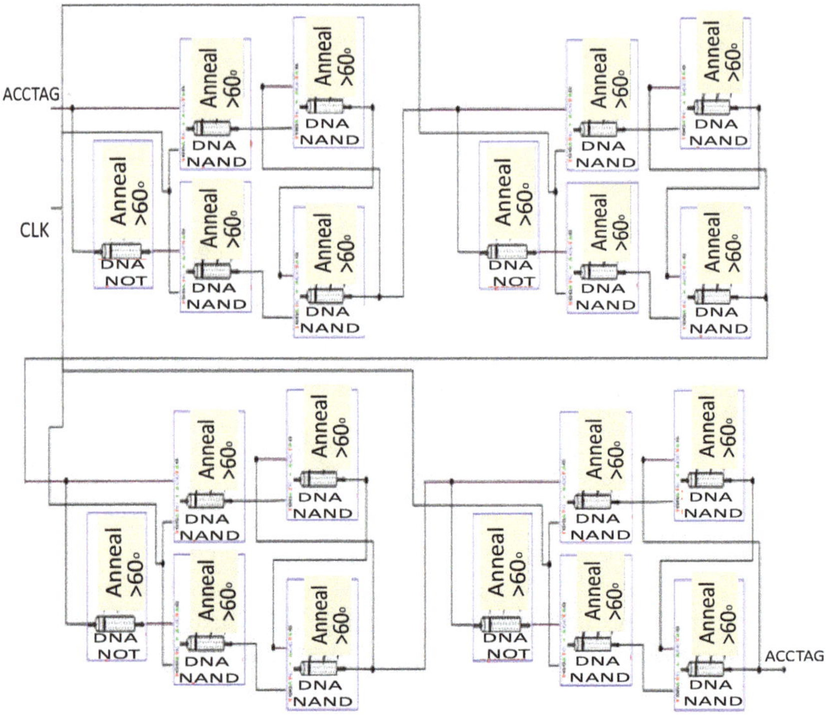

Figure 5.14.   DNA shift register.

The clock molecular input sequence is required for the DNA D flip-flop. It is seen from the diagram that the circuit has one molecular input sequence. One line of this molecular input sequence will be directed into a DNA NAND operation termed S molecular input sequence in Circuit. DNA NAND operation is performed here using S-molecular input sequences and Clock molecular input sequences.

X molecular input sequences another line, which performs a DNA not operation first. This DNA NOT operation was included into the DNA NAND operation with the designation R. The two molecular input sequences R and Clock are used in this DNA NAND operation. The output sequence of the DNA NAND operations is sent into the DNA SR latch as a molecular input sequence. These two molecular input sequences will be used to produce an SR latch. The output sequence of the DNA SR latch will be the final output sequences of

Q and $\overline{Q}$. Q will always produce the opposite of $\overline{Q}$. The DNA SR latch is discussed in this chapter titled DNA SR latch earlier in the book.

It produces one output sequence after processing the molecular input sequences in a DNA D flip-flop. The output sequence of the DNA D flip-flop is utilized as a clock molecular input sequence for the following DNA D flip-flop. As a result, a DNA shift register generates a single output molecular sequence.

5.7.3 Working principle

DNA shift registers are a kind of registers where both molecular sequence data loading, as well as data retrieval to/from the DNA shift register, occurs in serial mode sometimes. This research of synchronous DNA SISO shift register is sensitive to the positive edge of the clock pulse. Here the data word which is to be stored is fed sequence-by-sequence at the molecular input sequence of the first DNA flip-flop. Further, it is seen that the molecular input sequences of all other flip-flops are driven by the output sequences of the preceding ones say, for example, the molecular input sequence of DNA D flip-flop number-2 is driven by the output sequence of DNA D flip-flop number-1. At last, the data stored within the DNA register is obtained at the output sequence pin of the nth DNA D flip-flop in serial fashion.

Initially, all the DNA flip-flops in the DNA register are cleared by applying high on their clear pins. Next, the molecular input sequence data word is fed serially to DNA D flip-flop number-1.

This causes the molecular sequence appearing at the first pin to be stored into DNA D flip-flop number-1 as soon as the first leading edge of the clock appears. Further at the second clock tick, B1 gets stored into DNA D flip-flop number-2 while a new molecule enters into DNA flip-flop number-2.

This kind of shift in data continues for every rising edge of the clock pulse. This indicates that for every single clock pulse the data within the DNA register moves towards the right by a single molecule. Following the molecular sequence data transmission, as explained, one can note that the first molecular input sequence word appears at the output sequence of nth flip-flop for the nth clock tick. On applying further clock cycles, one gets the next successive molecular input sequence data word as the serial output sequence. Table 5.1 presents the truth table of a DNA D flip-flop.

After processing the molecular input sequences in a DNA D flip-flop, it creates one output sequence. The output sequence is used as a clock molecular input sequence for the next DNA D flip-flop. Thus, a DNA Shift Register creates one final output sequence.

5.7.4 *Applications*

The DNA shift register is a very much useful circuit in DNA computing. It can be used as DNA counter, DNA data format convertor, DNA data processor, etc.

5.7.4.1 *Data format converters of DNA shift registers*

Serial data transmission is preferred for long-distance communication due to its economic value in terms of the wires used. This necessitates parallel-to-serial conversion at the sender-end for which parallel-in serial-out DNA shift registers (DPISO) can be used. However, in general, many DNA microprocessor-based systems usually prefer a parallel form of data-in for which the transmitted molecule sequence data is to be converted into parallel mode using a serial-to-parallel converter like serial-in parallel-out DNA shift register (DSIPO).

5.7.4.2 *Counters of DNA shift registers*

DNA ring counter and the DNA Johnson counter are the two shift-register-based counters which are extensively used in digital applications. In DNA ring counters the output sequence of the last stage is back-fed as a molecular input sequence to the first stage. This causes the data stored within the DNA shift register to circulate within it continuously. DNA Johnson counter is similar to DNA ring counter except for the fact that the complement of the output sequence at the last stage of the DNA shift register is fed as a molecular input sequence to the first stage.

5.7.4.3 *Pseudo-random pattern generator of DNA shift registers*

DNA shift registers can be used to generate pseudo-random patterns which are used for testing. In order to achieve this, the output sequences of a few stages in the DNA shift register are XORed and connected as a molecular input sequence to the first stage of it.

The number of patterns generated depends upon the number of points that are tapped to be provided as DNA XOR operation molecular input sequences. If tapped appropriately, the maximum number of patterns that can be generated using an n-stage shift register is $(2n - 1)$.

DNA Shift register has some other applications also. DNA shift register can be used in various DNA devices. DNA shift register is fast more than a digital shift register so it will be more useful as well as can work simpler than today.

5.8   DNA Ripple Counter

A DNA counter is basically used to count the number of clock pulses applied to a DNA flip-flop. It can also be used for DNA frequency divider, DNA time measurement, DNA frequency measurement, DNA distance measurement, and also for generating square waveforms. In this, the DNA flip-flops are DNA asynchronous counters and are supplied with different clock signals, there may be a delay in producing the output sequence. Also, a few numbers of DNA logic operations are needed to design asynchronous counters. So, they are elementary in design and also are less expensive.

An n-molecular ripple counter can count up to $2n$ states. It is also known as MOD n counter. It is known as a ripple counter because of the way the clock pulse ripples its way through the DNA flip-flops. It is an asynchronous counter. Different DNA flip-flops are used with a different clock pulse. All the DNA flip-flops are used in toggle mode. Only one DNA flip-flop is applied with an external clock pulse and another flip-flop clock is obtained from the output sequence of the previous DNA flip-flop. The DNA flip-flop applied with an external clock pulse acts as LSB (least significant DNA sequence) in the counting sequence. A counter may be an up counter that counts upwards or can be a down counter that counts downwards or can do both, i.e. Count up as well as count downwards depending on the molecular input sequence control. The sequence of counting usually gets repeated after a limit.

DNA ripple counter is made out of four DNA JK flip-flops. Using these DNA JK flip-flops, the DNA ripple counter creates 4-molecular output sequences. Here in JK flip-flop, J and K are not shortened abbreviated letters of other words, such as "S" for Set and "R" for Reset, but are autonomous letters chosen by its inventor Jack Kilby to distinguish the flip-flop design from other types. Though Jack Kilby invented the digital electronics JK flip-flop. DNA ripple counter is an asynchronous counter. It is created using DNA JK flip-flops and these flip-flops are only controlled by clock pulse molecular input sequence.

DNA ripple counter produces much heat to produce the molecule's superposition state and also produces some garbage value.

5.8.1 *Block diagram*

DNA ripple counter uses four DNA JK flip-flops to create 4-molecular output sequences. DNA JK flip-flop has 2-molecular input sequence named as J and K. The block diagram of DNA ripple counter is shown in Figure 5.15.

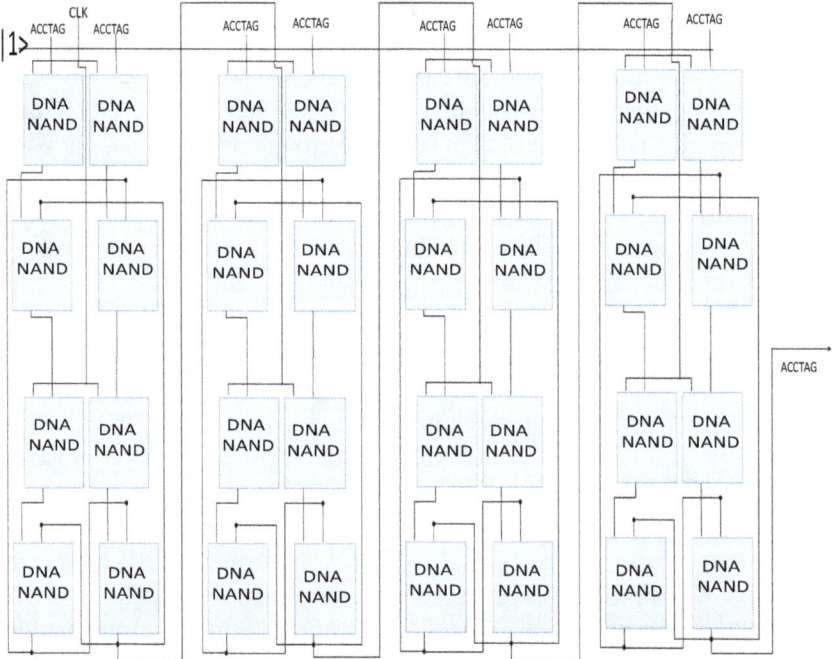

Figure 5.15. Block diagram of DNA ripple counter.

DNA JK flip-flop has the molecular input sequence J and K. This DNA JK flip-flop consists of many DNA NAND operations. At first basic DNA NAND operation performs a couple of then this operation output sequence is entered into the SR flip-flop as well as got the output sequence Q and $\overline{Q}$. First of all J and clk molecular input sequence perform the DNA NAND operation. The output sequence of the DNA NAND operation and output sequence of the DNA JK flip-flop $\overline{Q}$ performs another DNA NAND operation and produces the output sequence named as S. clk molecular input sequence is shared so K and clk molecular input sequence also performs the DNA NAND operation. This DNA NAND operations output sequence and Q output sequence of the DNA JK flip-flop performs another DNA NAND operation as well as produces the output sequence named as R.

These S and R entered into the DNA SR flip-flop and produce two output sequences Q and $\overline{Q}$. In DNA SR flip-flop, it has four DNA NAND operations. S molecular input sequence and clk molecular input sequence perform DNA NAND operation as well as R molecular input sequence and shared clk molecular input sequence also performs the DNA NAND operation. These two DNA NAND operations output sequence entered the DNA SR latch as molecular input sequence. In DNA SR latches two Q and one molecular input sequence as well as $\overline{Q}$ and other molecular input sequences perform the DNA NAND operation. Finally, after all of these DNA operations, the DNA JK flip flops final output sequences Q and $\overline{Q}$ are obtained. DNA SR flip-flop and SR latch are described already.

After processing the molecular input sequences in a DNA JK flip-flop, the output sequence of the flip-flop is going to be stored as an output sequence of the DNA ripple counter. DNA ripple counter can be performed as an up counter and also a down counter. Here every DNA JK flip-flop output sequence Q enters another DNA JK flip-flop as a clock pulse. This clk will decide thus the DNA JK flip-flop operational circuit will perform or not. Every DNA JK flip-flop operational circuit will produce the final output sequence.

5.8.2 *Circuit architecture*

DNA ripple counter uses four DNA JK flip-flops to create four molecular output sequences. DNA JK flip-flop has a 2-molecular input

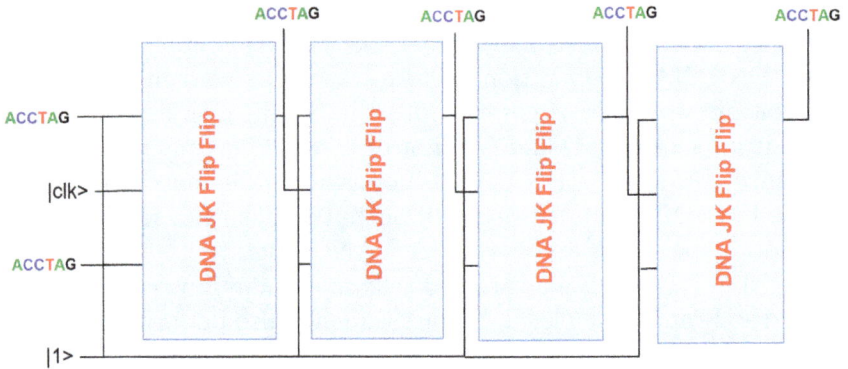

Figure 5.16.   Block diagram of DNA ripple counter.

sequence and one clock shared molecular input sequence. DNA JK flip-flops also depend on the clock. If the clock is enabled then the circuit will enable, otherwise not. Figure 5.16 presents the simplified block diagram of DNA ripple counter.

In DNA ripple counter there's a one clock molecular input sequence and one logic molecular input sequence which is shared in both J and K molecular input sequence port. Molecular input sequence J and molecular input sequence K both the value will perform the DNA NAND operation with the shared clk molecular input sequence differently. J and clk molecular input sequence perform the DNA NAND operation. A DNA NAND operation made by using DNA basic operations. For error correction here used an ancillary DNA sequence. In DNA NAND operation the value of an ancillary DNA sequence in ACCTAG. After J and clk perform the DNA NAND operation produced an output sequence and this output DNA sequence and the output sequence of the DNA flip-flop $\overline{Q}$ perform the DNA NAND operation and produce the output sequence S. Like in the same procedure K, clk and Q molecular input sequence produces the R output sequence. These operation circuits are mainly DNA NAND operations. This S and R molecular input sequence is performed in the DNA SR flip-flop operation. In DNA SR flip-flop operational circuit architecture also made by the proposed basic component of DNA computing is DNA NAND operation.

After processing the molecular input sequences in a DNA JK flip-flop, the output sequence of the DNA flip-flop is going to be stored as an output sequence of the DNA ripple counter. In DNA ripple counter, the intermediate architectures of four DNA JK flip-flops are connected in serial connection using the logical molecular input sequence. Every clock molecular input sequence as clk enters in every molecular sequence and from 2nd DNA JK flip-flop clk molecular input sequence is previous DNA Jk flip-flops first output sequence Q. With the same architecture DNA ripple counter can be performed as an up counter or as a down counter but clock pulse as clk need to sometimes have a positive edge triggered and sometimes a negative edge triggered.

5.8.3   *Working principle*

In the DNA ripple counter, there are four DNA JK flip-flop operational circuits. These DNA JK flip-flop operational circuits are connected in serial connection. In this DNA ripple counter, there is one clock molecular input sequence and a logic molecular input sequence which shared into the port J and K of DNA JK flip-flop operational circuit. The molecular input sequences J and K of a DNA JK flip-flop conducts two DNA processes in parallel. The DNA NAND operation is performed using J and shared molecular input sequence clk. The DNA NAND operations are then completed, and one of the DNA JK flip-flop's output sequences executes the DNA NAND operation, producing S. The K and clk molecular input sequences are performed first in the DNA NAND operation. The output sequence of the DNA JK flip-flop Q, as well as the result of the DNA first NAND operation, are then used to perform another DNA NAND operation, yielding R. The steps for creating S and R are carried out simultaneously. It is known that one of the distinctive properties of DNA operations is that they may do several operations at the same time, and this is exactly what is happening. The DNA SR flip-flop operation is then conducted on these S and R molecular input sequences. The DNA NAND operation, which is employed here, is also used to make DNA SR flip-flops. Two output sequences are discovered after conducting the DNA SR flip-flop operation. The opposite of one output sequence is the opposite of the other.

Here if the clk is TGGATC then the DNA JK flip-flop will not be triggered but if clk is ACCTAG the DNA JK flip-flop will be triggered and it will toggle the output sequence. First of all DNA JK

Table 5.6.   Truth table of DNA ripple counter.

clk	Q_0	Q_1	Q_2	Q_3
TGGATC	TGGATC	TGGATC	TGGATC	TGGATC
ACCTAG	ACCTAG	ACCTAG	ACCTAG	ACCTAG
ACCTAG	TGGATC	TGGATC	TGGATC	TGGATC

flip-flop operational circuit getting clk value is ACCTAG and toggles the output sequence value from the previous state value. Then the output sequence of the initial DNA JK flip-flop will be clk molecular input sequence of next DNA JK flip-flop. If the clk value is ACCTAG then the output sequence value will be toggled, otherwise the output sequence will be the previous state output sequence. Maintaining the same procedure every DNA JK flip-flop operated in the DNA ripple counter. DNA JK flip-flop is toggled very much, that's why in the DNA ripple counter DNA JK flip-flop operational circuit here is used as a basic component. The truth table of DNA ripple counter is presented in Table 5.6.

5.8.4 *Applications*

The counter in which the external clock is only given to the first flip-flop and the succeeding flip-flops are clocked by the output of the preceding flip-flop is called the asynchronous counter or ripple counter. The name ripple counter is because the clock signal ripples its way from the first stage of flip-flops to the last stage.

DNA binary coded decimal (BCD) counter is a decade counter which has Mod = 10. Mod means the number of states the counter has. DNA BCD counter counts decimal numbers from 0 to 9 and resets back to default 0. With each clock pulse, the counter counts up a decimal number. DNA ripple BCD counter is the same as DNA ripple up-counter, the only difference is when the DNA BCD counter reaches count 10 it resets its flip-flops. Block diagram of DNA BCD counter is given in Figure 5.17.

Different types of flip-flops with different clock pulses are used as a DNA ripple counter. It is an example of an asynchronous counter. The flip-flops are used in toggle mode in a DNA ripple counter. The external clock pulse is applied to only one flip-flop. The output of

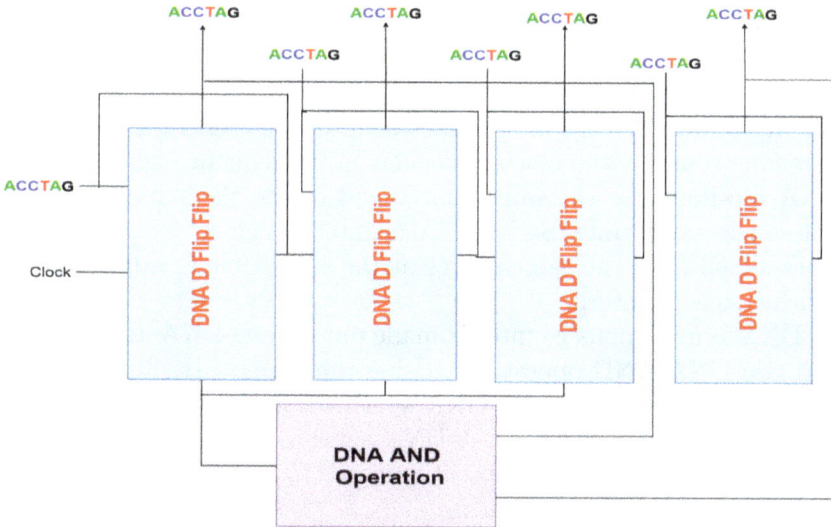

Figure 5.17. Block diagram of DNA BCD counter.

this flip-flop is treated as a clock pulse for the next flip-flop in the DNA ripple counter. In the counting sequence, the flip-flop in which external clock pulse is passed acts as LSB in DNA ripple counter operational circuits.

5.9 DNA Synchronous Counter

A DNA counter is a DNA device that can count any particular event on the basis of how many times the particular event(s) has occurred. In a DNA logic system or computer, this DNA counter can count and store the number of times any particular event or process has occurred, depending on a DNA clock signal. Most common type of DNA counter is a sequential DNA logic circuit with a single clock molecular input sequence and multiple output sequences. The output sequences represent two-valued decimal numbers. Each clock pulse either increases the number or decreases the number.

DNA synchronous circuit generally refers to something which is coordinated with others based on time. DNA synchronous signals occur at the same clock rate and all the clocks follow the same reference clock. DNA asynchronous counter has shown that the output sequence of that DNA counter is directly connected to the molecular input sequence of the next subsequent counter and making a chain system, and due to this chain system propagation delay appears during the counting stage and creates counting delays. In a DNA synchronous counter, the clock molecular input sequence across all the DNA flip-flops use the same source and create the same clock signal at the same time. So, a DNA counter which is using the same clock signal from the same source at the same time is called a DNA synchronous counter.

DNA synchronous counter is made out of four DNA JK flip-flops and two DNA AND operations. Using these DNA JK flip-flops and DNA AND operations, the DNA synchronous counter creates four molecular output sequences. A DNA synchronous counter produces much heat and this circuit operation needs to happen in the required environment of DNA computing.

DNA synchronous counter uses four DNA JK flip-flops to create four molecular output sequences. DNA JK flip-flop has the 2-molecular input sequence named as $|J\rangle$ and $|K\rangle$. The Details of DNA JK flip-flop have been discussed earlier in this chapter.

5.9.1 *Circuit architecture*

DNA synchronous counter uses four DNA JK flip-flops to create four molecular output sequences. DNA JK flip-flop has a 2-molecular input sequence and one clock shared molecular input sequence. DNA JK flip-flops also depend on the clock. If the clock is enabled then the circuit will be enabled, otherwise not. The simplified block diagram of DNA synchronous counter is shown in Figure 5.18.

Molecular input sequence J and molecular input sequence K both the value will perform the DNA NAND operation with the shared clk molecular input sequence differently. J and clk molecular input sequence performs the DNA NAND operation. DNA NAND operation is made by using DNA basic operations. In DNA NAND operation the value of an ancillary DNA sequence is ACCTAG. After J and clk perform the DNA NAND operation, an output sequence is created and this output sequence and the output sequence of the flip-flop $\overline{Q}$ performs the DNA NAND operation and produce the output sequence S. Like in the same procedure K, clk, and Q molecular sequences produce the R output sequence. These operational circuits

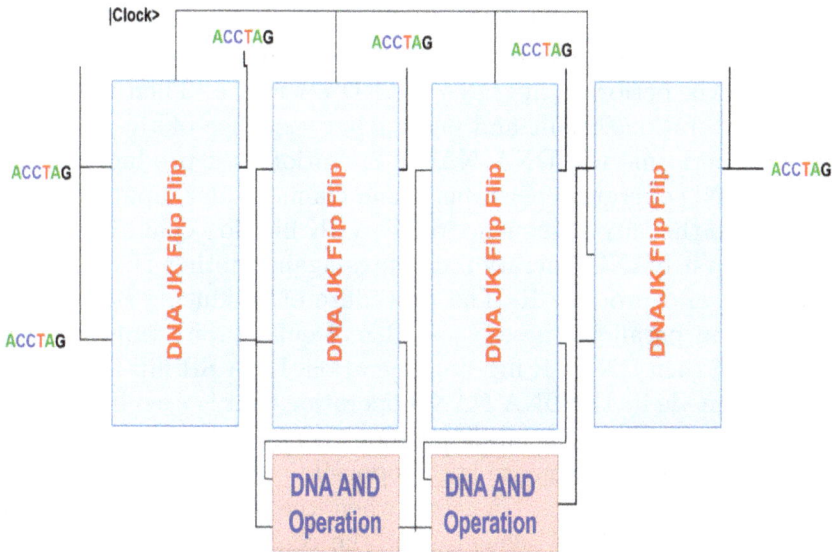

Figure 5.18.   Block diagram of DNA synchronous counter.

are mainly DNA NAND operations. This S and R molecular input sequence is performed in the DNA SR flip-flop operation. In DNA SR flip-flop operation circuit architecture also made by the proposed basic component of DNA computing is DNA NAND operation.

After processing the molecular input sequences in a DNA JK flip-flop, the output sequence of the DNA flip-flop is going to be stored as an output sequence of the DNA synchronous counter. Thus, DNA Synchronous Counter creates four molecular output sequences using four DNA JK flip-flops. The clock molecular input sequences for all of the four DNA JK flip-flops come from the same source. For this, all of the DNA flip-flops work synchronously. One output sequence of each of the second and third flip-flops go through DNA AND operations.

5.9.2   *Working principle*

DNA synchronous counter has one logical molecular shared input sequence and one clock molecular input sequence. DNA synchronous counter is constructed by the basic component as a DNA JK flip-flop operational circuit. DNA JK flip-flop is the 2-molecular circuit. DNA JK flip-flop is mostly used flip-flop in DNA computing.

DNA JK flip-flop's molecular input sequences J and K perform two DNA operations in parallelism. J and shared molecular input sequence clk performs the DNA NAND operation. Then this DNA NAND operations result and one output sequence of the DNA JK flip-flop performs the DNA NAND operation and produce S. Like DNA NAND operation performs K and clk molecular input sequence first. Then the output sequence of DNA JK flip-flop Q and the result of DNA first NAND operation executes again another DNA NAND operation and produce R. The procedure of producing S and R is executed in parallel. These S and R molecular input sequences are performed then DNA SR flip-flop operation. DNA SR flip-flop is also basically made by the DNA NAND operation that is used here. After performing the DNA SR flip-flop operation two output sequences are found. One output sequence is the opposite of another.

DNA JK flip-flop is triggered when the value of the clock is ACCTAG. So, in DNA synchronous the counter clock needs to be always high. According to the principle counters need to be toggled, that's why DNA JK flip-flop is perfect for DNA synchronous counters. DNA JK flip-flop operational circuit triggered then output sequence of DNA JK flip-flop will be shared molecular input sequence of second DNA JK flip-flop operational circuit. Then first DNA JK flip-flops output sequence and second DNA JK flip-flops output sequence enter the DNA AND operation circuit and perform the DNA AND operation. The produced output sequence from DNA AND operation performs the DNA JK flip-flop operation. Then again produced output sequence from the first DNA AND operation and the third DNA JK flip-flop operation's output sequence performs another DNA AND operation. Hence the previous DNA AND operations output sequence is shared into the DNA JK flip-flops as two molecular input sequences as well as performs the DNA JK flip-flop operational circuit. DNA synchronous counter circuit mainly performs as a finite counter. Table 5.7 represents the truth table of DNA synchronous counter.

5.9.3 *Applications*

As the name implies, the DNA synchronous counter contains flip-flops which are all in sync with each other, i.e. their clock molecules sequence inputs are connected together and are triggered by the same

Table 5.7. Truth table of DNA synchronous counter.

clk	Q_3	Q_2	Q_1	Q_0
TGGATC	TGGATC	TGGATC	TGGATC	TGGATC
ACCTAG	TGGATC	TGGATC	TGGATC	ACCTAG
ACCTAG	TGGATC	TGGATC	ACCTAG	TGGATC
ACCTAG	TGGATC	TGGATC	ACCTAG	ACCTAG
ACCTAG	TGGATC	ACCTAG	TGGATC	TGGATC
ACCTAG	TGGATC	ACCTAG	TGGATC	ACCTAG
ACCTAG	TGGATC	ACCTAG	ACCTAG	TGGATC
ACCTAG	TGGATC	ACCTAG	ACCTAG	ACCTAG
ACCTAG	ACCTAG	TGGATC	TGGATC	TGGATC
ACCTAG	ACCTAG	TGGATC	TGGATC	ACCTAG
ACCTAG	ACCTAG	TGGATC	ACCTAG	TGGATC
ACCTAG	ACCTAG	TGGATC	ACCTAG	ACCTAG
ACCTAG	ACCTAG	ACCTAG	TGGATC	TGGATC
ACCTAG	ACCTAG	ACCTAG	TGGATC	ACCTAG
ACCTAG	ACCTAG	ACCTAG	ACCTAG	TGGATC
ACCTAG	ACCTAG	ACCTAG	ACCTAG	ACCTAG
ACCTAG	TGGATC	TGGATC	TGGATC	TGGATC

external clock signal. This implies that all the DNA flip-flops update their values at the same time.

As the name suggests, DNA synchronous counters perform "counting" such as time and electronic pulses (external source like infrared light). They are widely used in lots of other designs as well such as DNA computing processors, DNA calculators, real time clocks, etc.

Alarm clock, set AC timer, set time in camera to take the picture, flashing light indicator in automobiles, car parking control, etc. can be constructed using a DNA synchronous counter.

This AC timer can be constructed using the DNA synchronous counter. Counting the time allotted for a special process or event by the scheduler can be made using a DNA synchronous counter. The UP/DOWN counter can be used as a self-reversing counter.

It is also used as a DNA clock divider circuit. The parallel load feature can be used to preset the counter for some initial count. Commons used in home appliances like washing machines, microwave ovens, Time schedule LED indicators, keyboard controllers, etc. can also be constructed using the DNA synchronous counter. They are also used in machine moving control. Most possible use will be in DNA multiplexing circuits. They are going to be used to generate saw-tooth waveforms.

5.10　Summary

For researchers, DNA computing is also a new field. Many different sorts of research are now being conducted. Sequential circuits in DNA computing mode were discussed in the preceding chapter. The same sequential circuits in DNA computing are introduced in this chapter. For all that DNA sequential circuits are discussed in a traditional fashion, this is a novel field. To construct sequential circuits, all essential DNA operations are utilized. Their connections and truth tables are displayed conventionally, but all of the components were in DNA computing mode. Heat conductance circuits are employed correctly when they are required. DNA circuits require more heat, which is delivered, and the entire method is demonstrated here to provide a comprehensive understanding of the DNA sequential circuit. Sequential circuits are not new to most people, but DNA sequential circuits are novel notions for scientists and researchers today.

Bibliography

Leonard M. Adleman. Molecular computation of solutions to combinatorial problems. *Science*, 266(5187): 1021–1024, 1994.

David H. Mathews, Jeffrey Sabina, Michael Zuker, and Douglas H. Turner. Expanded sequence dependence of thermodynamic parameters improves prediction of RNA secondary structure. *Journal of Molecular Biology*, 288(5): 911–940, 1999.

L.M. Smith, J.Z. Sanders, R.J. Kaiser, P. Hughes, C. Dodd, and C.R. Connell. C. Heiner, sb 11. kent. and le hood. *Nature*, 321(674): 7679, 1986.

Xuedong Zheng, Jing Yang, Changjun Zhou, Cheng Zhang, Qiang Zhang, and Xiaopeng Wei. Allosteric DNAzyme-based DNA logic circuit: Operations and dynamic analysis. *Nucleic Acids Research*, 47(3): 1097–1109, 2019.

Chapter 6

DNA Memory Devices

6.1 Introduction

Humans, by their nature, are archivists. Since the beginning of the digital age, vast troves of digital data have been hoarded. Multinational technology conglomerate Cisco predicted that by 2019, the data that makes up the "Internet of Everything," (often referred to as the Internet of Things), will surpass a seemingly unfathomable 500 zettabytes (500 sextillion bytes). A data volume that is rapidly inflating due to the wide adoption of personal cloud storage. However, the world's data storage capacity is not infinite as it relies on the element silicon which is rarely found pure. By tracking the progression of improvements to data storage density alongside the increase in data stored, researchers predict that by 2040 people will run out of memory-grade silicon on earth. An alternative to hard drives is progressing: DNA-based data storage. Deoxyribose nucleic acid (DNA) computing has the features of parallel processing and large storage capability that make it special from other conventional computing systems. It is a type of bimolecular programming where different types of reactions are used to perform basic operations and the processing information is stored in nucleic acids and proteins. DNA which consists of long chains of the nucleotides A, T, C, and G is life's information-storage material. Data can be stored in the sequence of these letters, turning DNA into a new form of information technology. It is already routinely sequenced (read), synthesized (written to) and accurately copied with ease. DNA is also incredibly stable, as has been demonstrated by the complete genome sequencing of a fossil horse that lived more than 500,000 years ago. And storing it does not require much energy. But it is the storage capacity that shines. DNA can accurately stow massive amounts of data at a density far exceeding that of electronic devices. The simple bacterium *Escherichia coli*, for instance, has a storage density of about 10^{19} DNA sequences per cubic centimeter, according to calculations published in 2016 in *Nature Materials* by George Church of Harvard University and his colleagues. At that density, all the world's current storage needs for a year could be well met by a cube of DNA measuring about 1 m aside. Among the challenges to making DNA data storage commonplace are the costs and speed of reading and writing DNA, which need to drop even further if the approach is to compete with electronic storage. Even if DNA does not become a ubiquitous storage material, it will almost certainly be used for generating information at entirely new scales and preserving certain types of data over the long term. This chapter presents the details of the same four memory devices shown in previous chapters but in DNA computing. That means DNA RAM, DNA ROM, DNA PROM, and DNA cache memory are described in detail.

6.2 DNA Random Access Memory

DNA memory storage will revolutionize the ability to store archival information on the Exabyte scale, not met by any existing archival memory storage technology. Chemically, DNA is an ideal long-term storage solution as it is an incredibly robust molecule, maintaining its integrity for tens of thousands of years when dried out. With current data storage architectures, the highest achieved storage density for synthetic DNA is a massive 215 petabytes per gram.

DNA random access memory (RAM), known as DNA RAM is a very important component of any computer. It stores important information for short-term use to make it easy for users to access the files so that the computer can write or read quickly. Unfortunately, any program data stored in RAM is lost when the device shuts down, and therefore, always save all information on hard drives to avoid loss. This section will describe the DNA RAM.

6.2.1 *History*

The use of DNA as a data storage medium has been the topic of a partnership between the University of Washington, Microsoft Research, and Twist Bioscience, which was revealed in spring 2016. Nature's technique of storing information about life's structure and function is through DNA. Life's data storage medium, with Twist Bioscience's high-throughput, high-quality bespoke DNA synthesis and the commonality of high-throughput next-generation sequencing may theoretically be modified to store digital data as well. The Universal Declaration of Human Rights, translated into 100 languages, and a music video by the band OK Go were among the materials saved in microscopic DNA sequences of DNA in 2016. In 2017, the collaboration announced the first report of archival quality data stored in DNA, preserving two recordings from the Montreux jazz festival which are part of UNESCO's Memory of the World Archive. February 2018 marks another huge step in this burgeoning industry. Microsoft and Washington researchers published the first report of DNA-based RAM in the journal *Nature*.

6.2.2 *Basic definition*

Methods of storing, processing, and selectively retrieving information contained inside sequence-controlled polymer barcoded nanoparticles are included in the DNA-based storage innovation. DNA memory storage is implemented by the inventors in DNA origami, structured DNA objects, and encapsulated DNA memory, which are physically linked together into memory blocks with unique ID tags. This enables the quick recovery of large-scale memory blocks utilizing associative memory and DNA logic, lowering the read-out time required for DNA sequencing by orders of magnitude while also allowing the data to be archived. Optical barcoding of DNA memory packets on DNA structures or encapsulated DNA memory enables molecularly encoded information to be sorted at speeds of up to 100 Mb/s. The structure of 2^k-to-n DNA RAM is shown in Figure 6.1.

The n data input lines provide the information to be stored in memory, and the n data output lines supply the information coming out of a particular word chosen among the 2^k available inside the memory.

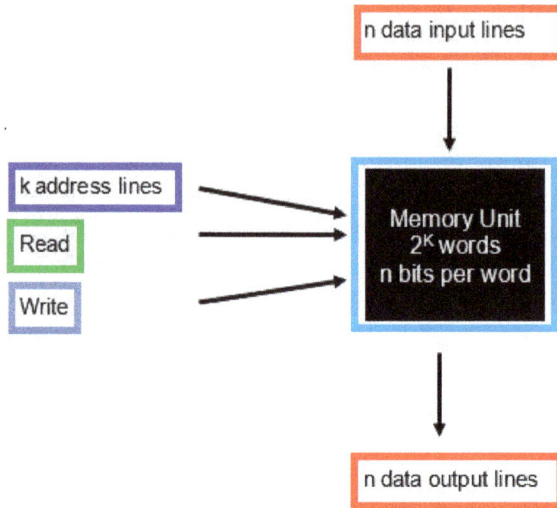

Figure 6.1. 2^k-to-n RAM.

6.2.3 *Advantages*

The advantages of DNA-based RAM are given as follows:

1. Massive storage capacity;
2. Very little energy is required for maintenance;
3. Rapid retrieval of random access memory including algorithmic sorting;
4. Rapid retrieval of large-scale memory.

6.2.4 *Disadvantages*

The major drawback of DNA-based data storage systems is the cost involved in writing and reading data on nucleotide sequences. The cost of synthesizing DNA (writing/encoding) is higher than that of sequencing (reading/decoding) also the processing speed is slow.

6.2.5 *Basic functions*

The major function of RAM is the provision of quick write and read access to a storage device. User's laptop loads information using RAM because it's quicker than a hard drive. The function of RAM can be equated to an office desk which is used to quickly access important writing tools and documents when the user needs them right away. Any information that's active on laptop, tablet, or Smartphone is stored temporarily in RAM which provides faster write/read times than hard drives. There are many basic and main characteristics of computer memory RAM, which are given as follows:

1. Long life;
2. Large size;
3. High power consumption;
4. Expensive;
5. No need to refresh;
6. Used as cache memory.

6.2.6 Block diagram

Figure 6.2 represents the block diagram of a 4-to-1 DNA RAM. This DNA RAM consists of four separate "Words" of memory and each is single sequenced wide. The DNA RAM cell has three input sequences and one output sequence.

6.2.7 Design architecture of basic components

A DNA RAM consists of three basic components such as DNA decoder, molecular cells, and DNA OR operations. To execute a 4-to-2 DNA RAM operation, it is necessary to get the following:

1. A 2-to-4 DNA decoder,
2. Molecular cells, and
3. DNA OR operations for corresponding minterms are required.

DNA decoder and DNA OR operations are discussed earlier.

6.2.7.1 Circuit design of molecular cell

The fundamental design of this Molecular cell is based on the DNA R-S flip-flop (Figure 6.3). To begin with, the molecular cell has three inputs and a single output. The inputs are labeled "Select," "R/W,"

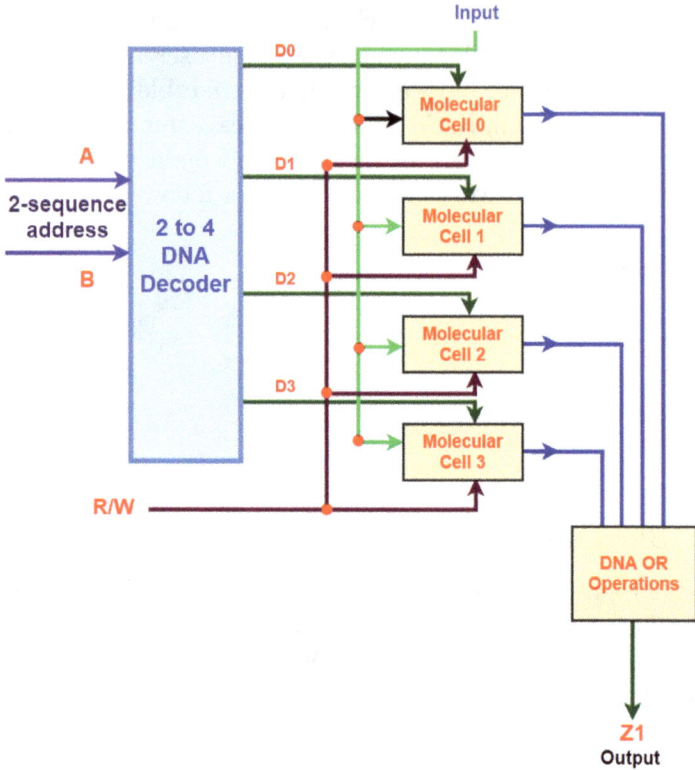

Figure 6.2.   Block diagram of a 4-to-1 DNA RAM.

and "Input." The output line is labeled "M0." To perform the Molecular cell output, two DNA NOT, six DNA AND, and two DNA NOR operations are needed to perform.

Step 1: First draw three input sequences Input, R/W, and Select. Two possible states for a sequence are the states "**TGGATC**" false, and "**ACCTAG**" true.

Step 2: Draw DNA NOT operation with the input and R/W sequences.

Step 3: Each DNA AND operation has three inputs. So two inputs are taken (NOT input and select) to one DNA AND operation input and the output of this operation will go to another DNA AND operation with input sequence R/W as input.

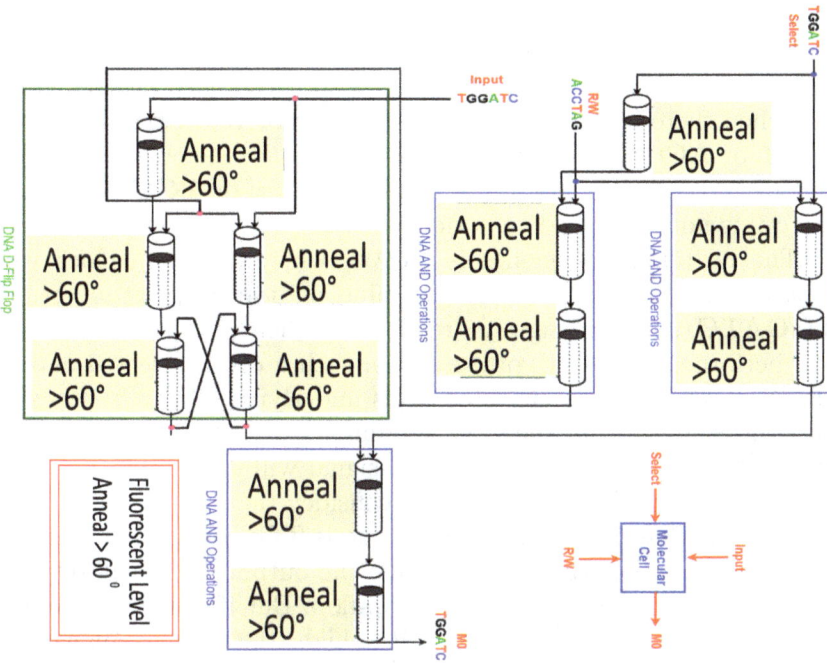

Figure 6.3. DNA single molecular cell.

Step 4: Again, each DNA AND operation has three inputs. So two inputs are taken (input and select) to one DNA AND operation input and the output of this operation will go to another DNA AND operation with input sequence R/W as input.

Step 5: The outputs of Steps 3 and 4 will go to the DNA R-S flip-flop as input.

Step 6: Finally, the output of DNA R-S flip-flop and select will go to a DNA AND operation, and then this output with DNA NOT operation of R/W will go through another DNA AND operation to produce the desired molecular cell output.

6.2.7.2   *Working procedure of molecular cell*

In sequential devices as simple as a DNA R-S flip-flop could be used to remember one DNA sequence of data. To develop a complete memory cell, called a sequence cell, based on the flip-flop. The number of total DNA cells per word will be $m \times n$, where m represents groups with n DNA sequences. The "select" input is used to access the cell, either for reading or writing also used to access any one sequence cell when there is more than one sequence cell. When the select line is high or **ACCTAG** then the cell performs the memory operation. But when the select line of the DNA cell is low or **TGGATC** then the cell is not interested to perform a read from or written to.

The next input sequence is "R/W" where a system clock will conduct this input. If the clock value on the read/write line is **TGGATC**, this will signify "read" and when it is **ACCTAG**, it will perform the "write" phase. When such a cell is selected and in "read" mode, the current value of its underlying DNA flip-flop will be transferred to the cell's output line. When the cell is selected and in "write" mode, an input data signal will determine the value remembered by the flip-flop. Let's consider the cell that has been selected. In that case, if the clock value is **TGGATC** then the cell contents are to be read and this time the output value will depend only on the P-valued of the DNA flip-flop. But if P is low, the cell output will be **TGGATC** and if P is high, the cell output will be **ACCTAG**. It occurs because the DNA AND operation added to the cell's output which has three input-negated R/W, Select, and P; and both "negated R/W" and "Select" is currently high **ACCTAG**. Figure 6.4 shows the four-sequence molecular cells to perform M0 to M3 for further minterms operation.

To perform the 4-to-1 DNA RAM, four selection lines are needed to design four DNA sequence cells. The output of DNA 2-to-4 decoder with four output sequences will perform as selection input sequence of sequence cell.

6.2.8 *Circuit architecture of 4-to-1 DNA RAM*

In 4-to-1 DNA RAM architecture (Figure 6.5), two address lines are needed with 4-to-1 DNA sequence RAM, and each address line needs to be DNA NOT form as well. These address lines combination will be the input of 2-to-4 DNA decoders which consists of four DNA AND operations and this decoder has one enable input. Four select lines are obtained from this decoder and each select line will go through each molecular cell. Note that the word calculation of RAM will be 2^k, where k is the address line and 2^k is the total words of n DNA

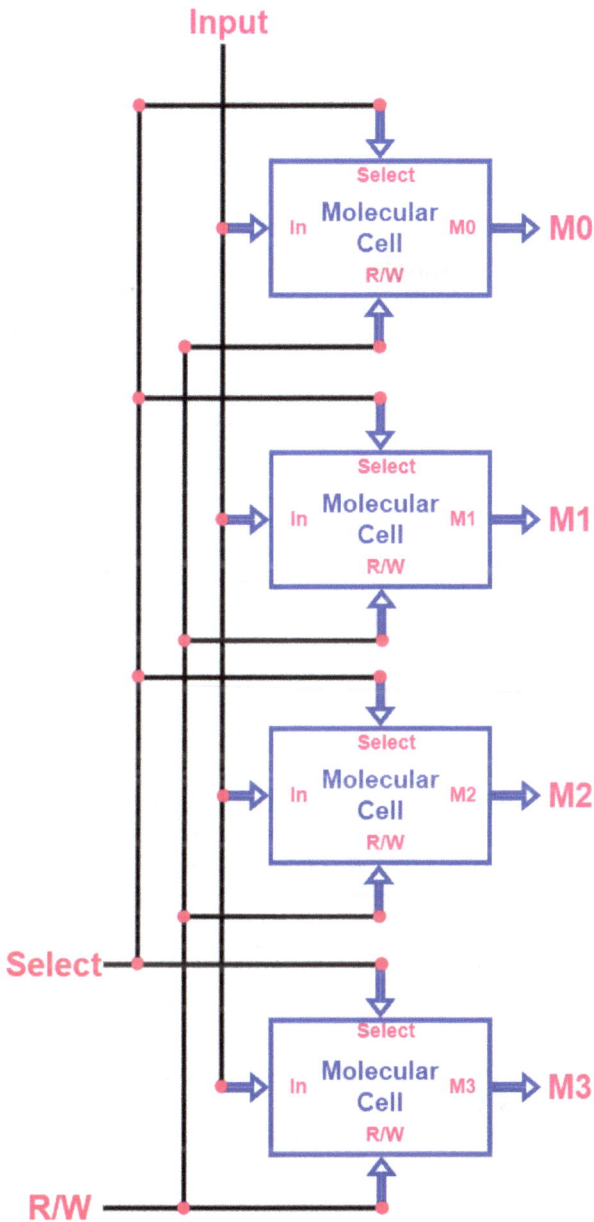

Figure 6.4.   DNA molecular cells.

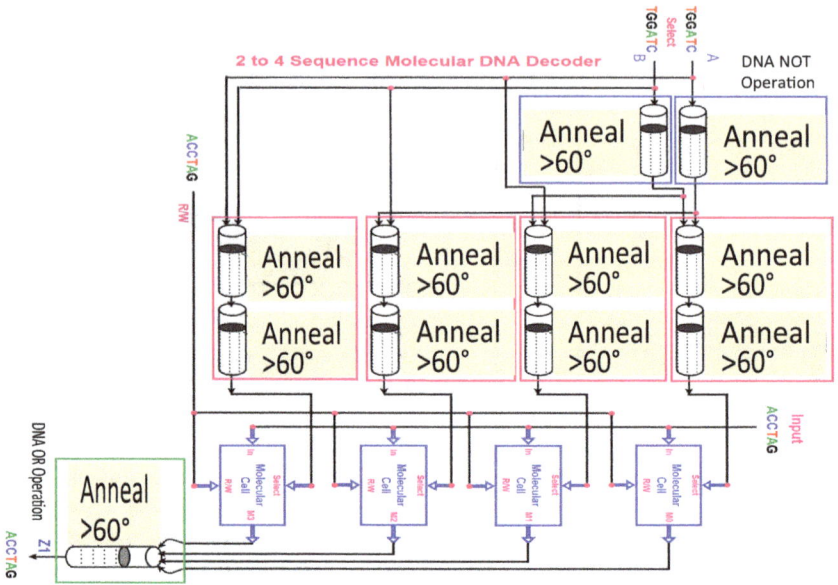

Figure 6.5.   4-to-1 DNA RAM.

Table 6.1.   Control input to memory chip.

R/W	Memory operation
X	*None*
TGGATC	*Write to selected word*
ACCTAG	*Read from selected word*

sequences and the decoder combination will be $k \times 2^k$. This single sequence RAM consists of four separate molecular cells and each cell has three inputs — D0, anyone selects line and read/write inputs. The obtained output from four DNA molecular RAM cells will be the input of a DNA OR operation which produces the final output. This is the whole design procedure of 4-to-1 DNA RAM. The control input to memory chip is given in Table 6.1.

6.2.9   *Working principle*

Figure 6.5 represents the implementation of 4-to-1 DNA RAM. This DNA RAM consists of four separate "Words" of memory and each is

1-molecular sequence wide. The DNA RAM Cell has three inputs and one output. The complete circuit of a DNA RAM cell is described in Figure 6.5 with proper explanation. A word consists of two DNA RAM cells arranged in such a way so that both sequences can be accessed simultaneously. Four words of memory need two address lines. A and B are the two-sequence address lines input that goes through a DNA 2-to-4 decoder that selects one of the four words. The memory-enabled input enables the decoder. If the memory enables is **TGGATC**, all output of the decoder will be **TGGATC** and in that case, none of the memory addresses will be selected. But when the memory enables is **ACCTAG** then one of the four words is selected. The word is selected by the value in the two address lines. When a word has been selected, the read/write input determines the operation. During the read operation, the four sequences of the selected word pass to the DNA OR operations to the output Z1 terminals. But during the write operation, the data which is available in the input lines are transferred into the four DNA cells of the selected word. The DNA RAM cells that are not selected are become disable and their previous sequence never changes. But when the memory enable input that passes into the decoder is equal to **TGGATC**, none of the words are selected, and then all DNA cells remain unchanged regardless of the value of the read/write input. This is the working procedure of 4-to-1 DNA RAM.

6.2.10 *Applications*

Sometimes, the contents of a relatively slow ROM chip are copied to read/write memory to allow for shorter access times. The ROM chip is then disabled while the initialized memory locations are switched in on the same block of addresses (often write-protected). This process is sometimes called shadowing, is fairly common in both computers and embedded systems.

As a common example, the BIOS in typical personal computers often has an option called "use shadow BIOS" or similar. When enabled, functions relying on data from the BIOS's ROM will instead use DRAM locations (most can also toggle shadowing of video card ROM or other ROM sections). Depending on the system, this may not result in increased performance and may cause incompatibilities. For example, some hardware may be inaccessible to the operating system if shadow RAM is used. On some systems, the benefit may be hypothetical because the BIOS is not used after booting in favor of direct hardware access. Free memory is reduced by the size of the shadowed ROMs.

6.3 DNA Read Only Memory

The traditional read only memory (ROM) is a slower memory. Thus, DNA computing enables the creation of new types of computers which is capable of operating sequences as input states by increasing storage capacity. DNA memory is required for the formation of a synchronization tool that can match the multiple procedures in a DNA-based computer, a DNA operation that retains the identity of any state, and a method for turning preset molecular sequences into on-demand sequences, among other DNA information processing devices. DNA-based memory may be utilized in a variety of applications, including DNA computing and DNA communication. Continuous research and experimentation have enabled the storing of sequences in DNA memory. DNA memory is the DNA-mechanical equivalent of conventional computer memory in DNA computing. Unlike conventional memory, which stores information as binary states (represented by "1"s and "0"s), DNA memory saves a molecular state for subsequent retrieval. Sequences (represented by "**ACCTAG as 1**" and "**TGGATC as 0**") which provide important computing information, are stored in these states. Unlike traditional computer memory, the states saved in DNA memory can be in a sequence, providing far more practical flexibility in DNA algorithms than traditional information storage.

6.3.1 *History*

The idea of DNA digital data storage dates back to 1959, when the physicist Richard P. Feynman, in *There's Plenty of Room at the Bottom: An Invitation to Enter a New Field of Physics* outlined the general prospects for the creation of artificial objects similar to objects of the microcosm (including biological) and having similar or even more extensive capabilities. In 1964–1965 Mikhail Samoilovich Neiman, the Soviet physicist, published three articles about micro miniaturization in electronics at the molecular-atomic level, which independently presented general considerations and some calculations regarding the possibility of recording, storage, and retrieval of information on synthesized DNA and RNA molecules. N. Wiener expressed ideas about the miniaturization of computer memory, close to the ideas, proposed by M. S. Neiman independently. One of the earliest uses of DNA storage occurred in a 1988 collaboration between artist Joe Davis and researchers from Harvard. The image, stored in a DNA sequence in *E. coli*, was organized in a 5 × 7 matrix that, once decoded, formed a picture of an ancient Germanic rune representing life and the female Earth. In the matrix, ones corresponded to dark pixels while zeros corresponded to light pixels. In 2007, a device was created at the University of Arizona using addressing molecules to encode mismatch sites within a DNA strand. These mismatches were then able to be read out by performing a restriction digest, thereby recovering the data. In 2011, George Church, Sri Kosuri, and Yuan Gao carried out an experiment that would encode a 659-kb book that was co-authored by Church. To do this, the research team did a two-to-one correspondence where a binary zero was represented by either adenine or cytosine and a binary one was represented by guanine or thymine. After examination, 22 errors were found in the DNA. In 2012, George Church and colleagues at Harvard University published an article in which DNA was encoded with digital information that included an HTML draft of a 53,400-word book written by the lead researcher, 11 JPG images, and 1 JavaScript program. Multiple copies for redundancy were added and 5.5 petabits can be stored in each cubic millimeter of DNA. The researchers used a simple code where DNA sequences were mapped one-to-one with bases, which had the shortcoming that it led to long runs of the same base, the sequencing of which is error-prone. This result showed that besides its other functions, DNA can also be another type of storage media such as hard drives and magnetic tapes. In 2013, an article led by researchers from the European Bioinformatics Institute (EBI) and submitted at around the same time as the paper of Church and colleagues detailed the storage, retrieval, and reproduction of over five million DNA sequences of data. All the DNA files reproduced the information between 99.99% and 100% accuracy. The main innovations in this research were the use of an error-correcting encoding scheme to ensure an extremely low data-loss rate, as well as the idea of encoding the data in a series of overlapping short oligonucleotides identifiable through a sequence-based indexing scheme. Also, the sequences of the individual strands of DNA overlapped in such a way that each region of data was repeated four times to avoid errors. Two of these four strands were constructed backward, also to eliminate errors. The costs per megabyte were estimated at \$12,400 to encode data and \$220 for retrieval. However, it was noted that the exponential decrease in DNA synthesis and sequencing costs, if it continues, should make the technology cost-effective for long-term data storage by 2023. In 2013, a software called DNACloud was developed by Manish K. Gupta and coworkers to encode computer files to their DNA representation. It implements a memory efficiency version of the algorithm proposed by Goldman *et al.* to encode (and decode) data to DNA (.dnac files). The long-term stability of data encoded in DNA was reported in February 2015, in an article by researchers from ETH Zurich. The team added redundancy via Reed–Solomon error correction coding and by encapsulating the DNA within silica glass spheres via sol–gel chemistry. In 2016, research by Church and Technicolor Research and Innovation were published in which, 22 MB of an MPEG compressed movie sequence were stored and recovered from DNA. The recovery of the sequence was found to have zero errors. In March 2017, Yaniv Erlich and Dina Zielinski of Columbia University and the New York Genome Center published a method known as DNA Fountain that stored data at a density of 215 petabytes per gram of DNA. The technique approaches the Shannon capacity of DNA storage, achieving 85% of the theoretical limit. The method was not ready for large-scale use, as it costs \$7000 to synthesize 2 megabytes of data and another \$2000 to read it. In March 2018, the University of Washington and Microsoft published results demonstrating storage and retrieval of approximately 200MB of data. The research also proposed and evaluated a method for random access of data items stored in DNA. In March 2019, the same team announced they had demonstrated a fully automated system to encode and decode data in DNA. Research published by Eurecom and Imperial College in January 2019, demonstrated the ability to store structured data in synthetic DNA. The research showed how to encode structured or, more specifically, relational data in synthetic DNA and also demonstrated how to perform data processing operations (similar to SQL) directly on the DNA as chemical processes.

The first article describing data storage on native DNA sequences via enzymatic nicking was published in April 2020. Here, scientists demonstrate a new method of recording information in the DNA backbone that enables sequence-wise random access and in-memory computing.

6.3.2 *Basic definition*

DROM is an abbreviation for DNA ROM. It refers to computer memory chips that hold permanent or semi-permanent data and incorporate both the DNA decoder and DNA OR operations onto a single integrated circuit (IC). The contents of DNA ROM are non-volatile; even if the computer is turned off, the contents of DNA ROM persist. It is used to hold a computer's boot-up instructions. Almost every computer has a tiny DNA sequence of DNA ROM that contains the boot software. This is made up of a few kilobytes of code that instructs the computer on what to do when it boots up, such as conducting hardware diagnostics and loading the operating system into DNA RAM. The BIOS is the boot firmware on a computer. To update the programming in DNA ROM, these chips have to be physically removed and replaced. Data saved in DNA ROM cannot be electrically changed once the memory device is manufactured. A block diagram of a ROM is shown in Figure 6.6. It consists of n input lines and m output lines. Each DNA sequence combination of the input variables is called an address. Each DNA sequence combination that comes out of the output lines is called a word. The number of DNA sequences per word is equal to the number of output lines m. An address is essentially a binary number that denotes one of the minterms of n variables. The block diagram 2^k-to-n DNA ROM is shown in Figure 6.6.

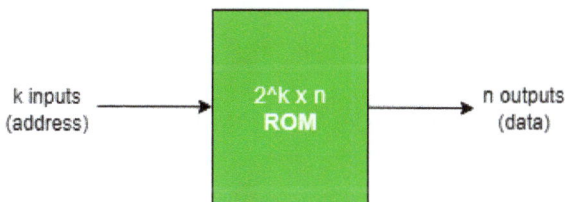

Figure 6.6.   Block diagram of 2^k-to-n DNA ROM.

Initially, the DNA ROM is a combinational circuit with DNA AND operations connected as a DNA decoder and several DNA OR operations equal to the outputs in the unit. Therefore it is a two-level implementation in the sum of minterms form. With k input lines and n output lines in ROM, the output functions will be calculated through the sum of minterms form. The number of distinct addresses possible with k-input variables is 2^k. An output word can be selected by a unique address, and since there are 2^k distinct addresses in a ROM, there are 2^k distinct words that are said to be stored in the unit. The word available at the output lines at any given time depends on the address value applied to the input lines. Therefore, A ROM is characterized by the number of words 2^k and the number of DNA sequences per word m. For input, $k = 2$ and output, $n = 2$ the ROM circuit will be called 4-to-2 ROM, and the function output F_1 and F_2 in the sum of minterms form, $\sum(0, 1, 2, 3)$. It does not have to be a DNA AND-OR implementation but it can be any other possible two-level minterms implementation. Thus the second level is usually a wired logic connection to facilitate the fusing of links.

6.3.3 *Advantages*

DNA ROM stores the instructions required for communication between various hardware components. As previously stated, it is required for the storage and functioning of the BIOS, but it may also be used for basic data management, to store software for basic utility functions, and to read and write to peripheral devices. Other benefits of DNA ROM include:

1. DNA ROM has more storage capacity.
2. More dependable than DNA RAM since it is non-volatile and cannot be altered or modified unintentionally.
3. Data can be stored permanently as DNA molecular sequences.

6.3.4 *Disadvantages*

There are several drawbacks of DNA ROM, which are listed as follows:

1. DNA ROM is read-only and cannot be altered. That is, DNA ROM Memory can only read data and cannot modify the DNA ROM data.

2. DNA ROM is a slower form of memory.
3. Improperly deleting the data of the DNA ROM Memory would brick the memory.
4. Some DNA ROM Memory types allow the user to rewrite the contents.

6.3.5 *Basic functions*

Information is stored in memory cells or memory locations in primary memory. Information that can be accessed is stored in memory locations. A single memory access operation must be performed to read or write data in a memory location, which necessitates the supply of independent control signals to the memory. A random memory enables unrestricted in space and time access to any location at any address in the address space. The access is possible independently on the order of all previous accesses. The access can take place to an address in any order. Each location in a random access memory has independent hardware circuits that provide the access. These circuits are activated as a result of address decoding. The DNA ROM parameters are: First, memory volume is the number of locations that exist in a given memory. When words are used, the length of a word in DNA sequences has to be given. Second, memory access time is the time that separates sending a memory access request and the reception of the requested information. The access time determines the unitary speed of a memory (the reception time of unitary data). The access time is small for fast memories. Third, memory cycle time is the shortest time that has to elapse between consecutive access requests to the same memory location. The memory cycle time is another parameter that characterizes the overall speed of the memory. The speed is big when the cycle time is small. Finally, the memory transfer rate is the speed of reading or writing data in the given memory, measured in DNA sequences/second.

6.3.6 *Block diagram*

The block diagram of 4-to-2 DNA ROM is shown in Figure 6.7, the unit consists of four words of two input sequences (A = **ACCTAG (1) or TGGATC (0)** and B = **ACCTAG (1) or TGGATC (0)**) each. There are only two input sequences in DNA 4-to-2 ROM because $2\$2^2 = 4$ and with two sequence variables can specify

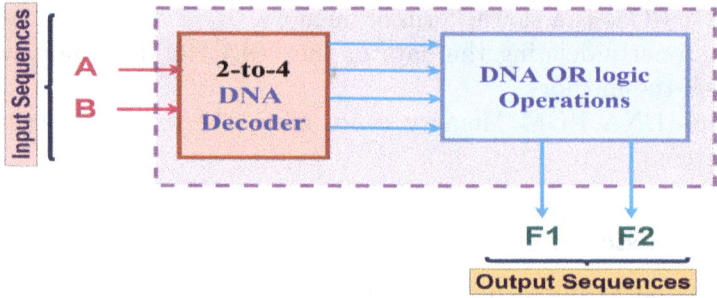

Figure 6.7.   Block diagram of 4-to-2 DNA ROM.

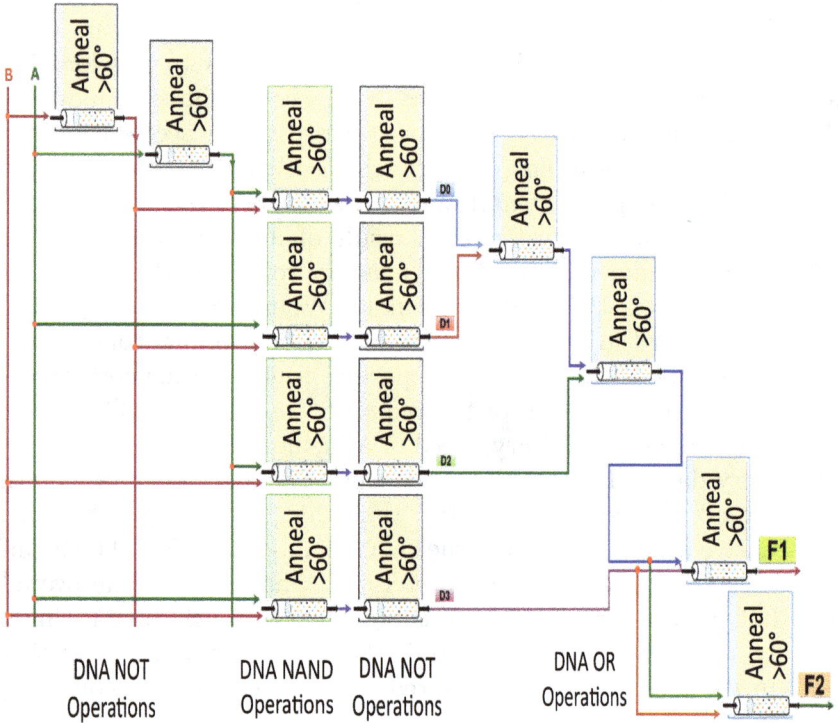

Figure 6.8.   4-to-2 DNA ROM.

four addresses. To perform minterms of four addresses, a DNA 2-to-4 decoder and DNA OR operations are required. Thus the input sequence address is **TGGATC**, **TGGATC**, word number 0 is selected and it appears on the output sequence lines. If the

Table 6.2. Truth table of 4-to-2 DNA ROM.

B	A	F1	F2
TGGATC	TGGATC	ACCTAG	ACCTAG
TGGATC	ACCTAG	ACCTAG	ACCTAG
ACCTAG	TGGATC	ACCTAG	ACCTAG
ACCTAG	ACCTAG	ACCTAG	ACCTAG

input sequence address is **ACCTAG, ACCTAG**, word number 3 is selected and it appears on the output lines. In between, there are two other input sequence addresses that can select other two words. DNA 4-to-2 ROM circuit architecture is shown in Figure 6.8.

6.3.7 *Working principle*

According to the truth table of DNA 4-to-2 ROM (Table 6.2) and also considering Table 6.1, it is necessary to do the following operations to perform desired output sequences:

(i) For input combination A, B = **TGGATC, TGGATC**, D0 line will be open. So, the value of D0 will be **ACCTAG** and D1 to D3 will be **TGGATC**. For the output of F1 and F2, perform DNA OR operations among D0 to D3 and generate **ACCTAG** as the desired output.

(ii) For input combination A, B = **ACCTAG, TGGATC**, D1 line will be open. So, the value of D1 will be **ACCTAG** and D0, D2 and D3 will be **TGGATC**. For the output of F1 and F2, perform DNA OR operations among D0 to D3 and generate **ACCTAG** as the desired output.

(iii) For input combination A, B = **TGGATC, ACCTAG**, D2 line will be open. So, the value of D2 will be **ACCTAG** and D0, D1, and D3 will be **TGGATC**. For the output of F1 and F2, perform DNA OR operations among D0 to D3 and generate **ACCTAG** as the desired output.

(iv) For input combination A, B = **ACCTAG, ACCTAG**, D3 line will be open. So, the value of D3 will be **ACCTAG** and D0 to D2 will be **TGGATC**. For the output of F1 and F2, perform DNA OR operations among D0 to D3 and generate **ACCTAG** as the desired output.

6.3.8   *Applications*

The main concern of DNA ROM application is data storage and memory technology. DNA ROMs are taken to include all semiconductor, nonvolatile memory devices, and they are used in applications where nonvolatile storage of information, data, or program codes is needed and where the stored data rarely or never change. Here are some of the most common application areas:

1. BIOS chip in computers;
2. Network operating systems;
3. Server operating systems;
4. Storing fonts for laser printers;
5. Storing sound data in electronic musical instruments; and
6. Storage for in-built self-learning functionality in remote-operated transmitters.

6.4   DNA Programmable Read-Only Memory

Advances in memory technology have helped make more advanced and compact devices possible. But electronic memory consumes quite a lot of power, can be difficult to read and write, and has a limited lifespan. But billions of years ago, nature came up with a system for storing information: DNA. The average human stores 40 exabytes–40 million terabytes of information in their DNA every day and DNA can potentially remain stable for thousands of years. The major concern is to create a DNA-based programmable read-only memory (PROM) that can be programmed, read electronically, and interface with electronic devices. For biological purposes, DNA stores information as a series of chemical bases, adenosine, cytosine, guanine, and thiamine, represented by the letters **A**, **C**, **G**, and **T**. This code is "read" by enzymes to make proteins.

DNA computing has the features of parallel processing and large storage capability that make it special PROM other conventional computing systems. It is a type of bimolecular programming where different types of reactions are used to perform basic operations and the processing information is stored in nucleic acids and proteins. The traditional PROM is a slower memory. Thus, DNA computing enables the creation of new types of computers which is capable of operating sequences as input states by increasing storage capacity. DNA memory is required for the formation of a synchronization tool that can match the multiple procedures in a DNA-based computer, a DNA operation that retains the identity of any state, and a method for turning preset molecular sequences into on-demand sequences, among other DNA information processing devices. DNA-based memory may be utilized in a variety of applications, including DNA computing and DNA communication. Continuous research and experimentation have enabled the storing of sequences in DNA memory. DNA memory is the DNA-mechanical equivalent of conventional computer memory in DNA computing. Unlike conventional memory, which stores information as binary states (represented by "1"s and "0"s), DNA memory saves a molecular state for subsequent retrieval. Sequences (represented by "**ACCTAG as 1**" and "**TGGATC as 0**"), which provide important computing information, are stored in these states. Unlike traditional computer memory, the states saved in DNA memory can be in a sequence, providing far more practical flexibility in DNA algorithms than traditional information storage.

6.4.1 *Basic definition*

DNA PROMs are used in digital electronic devices to store permanent data, usually low-level programs such as firmware or microcode. The key difference from a standard ROM is that the data is written into a ROM during manufacture, while with a PROM the data is programmed into them after manufacture.

It refers to DNA computer memory chips that hold permanent or semi-permanent data and incorporate both the DNA decoder and DNA OR operations onto a single IC. The contents of DNA PROM are non-volatile; even if the computer is turned off, the contents of DNA PROM persist. It is used to hold a computer's boot-up instructions. Almost every computer has a tiny nucleotide sequence of DNA PROM that contains the boot software. This is made up of a few kilobytes of code that instructs the computer on what to do when it boots up, such as conducting hardware diagnostics and loading the operating system into DNA PRAM. The BIOS is the boot firmware on a computer. To update the programming in DNA PROM, the DNA PROM chips have to be physically removed

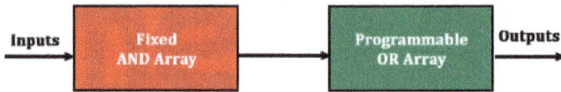

Figure 6.9.   Block diagram of DNA PROM.

and replaced. Data saved in DNA PROM cannot be electrically changed once the memory device is manufactured. A block diagram of a DNA PROM is shown in Figure 6.9. It consists of n input lines and m output lines. Each DNA sequence combination of the input variables is called an address. Each DNA sequence combination that comes out of the output lines is called a word. The number of DNA sequences per word is equal to the number of output lines m. An address is essentially a binary number that denotes one of the minterms of n variables.

Initially, the DNA PROM is a combinational circuit with DNA AND operations connected as a DNA decoder and several DNA OR operations equal to the outputs in the unit. Therefore it is a two-level implementation in the sum of minterms form. With n-input lines and m-output lines in DNA PROM, the output functions will be calculated through the sum of minterms form. The number of distinct addresses possible with n-input variables is 2^n. An output word can be selected by a unique address, and since there are 2^n distinct addresses in a DNA PROM, there are 2^n distinct words that are said to be stored in the unit. The word available at the output lines at any given time depends on the address value applied to the input lines. Therefore, with n-input lines and m-output lines in DNA programmable ROM, the output functions will calculate through the programmable sum of minterms form. For input, $n = 2$ and output, $m=2$ the DNA PROM circuit will be called 4-to-2 DNA PROM and the function output $F_1 = \sum(0,2)$ and $F_2 = \sum(1,3)$.

6.4.2   *Advantages*

DNA PROM stores the instructions required for communication between various hardware components. As previously stated, it is required for the storage and functioning of the BIOS, but it may also be used for basic data management, to store software for basic utility functions, and to read and write to peripheral devices. Other benefits of DNA PROM include:

1. Contents are always known and verifiable.
2. The programming can be done using many types of software and does not rely on the hard wiring of the program to the chip.
3. Since it is not possible to un-blow the fuse, the authenticity of the data remains intact and it is impossible to remove or alter the contents.

6.4.3 Disadvantages

There are several drawbacks of PROM, which are listed as follows:

1. The biggest disadvantage of PROM is that the data once burnt cannot be erased or changed when detected with errors.
2. DNA PROM is a slower form of memory.
3. Improperly deleting the data of the DNA PROM would brick the memory.

6.4.4 Basic functions

On the CPU and other computer equipment, memory devices are used to store various types of information such as data, programs, addresses, textual files, and status information. Information is stored in memory cells or memory locations in primary memory. Information that can be accessed is stored in memory locations. A single memory access operation must be performed to read or write data in a memory location, which necessitates the supply of independent control signals to the memory. Memory capacity or memory volume is the number of locations that exist in a given memory. Memory capacity is measured in DNA sequences, bytes, or words. When words are used, the length of a word in DNA sequences has to be given. Memory access time is the time that separates sending a memory access request and the reception of the requested information. The access time determines the unitary speed of memory (the reception time of unitary data). The access time is small for fast memories. Memory cycle time is the shortest time that has to elapse between consecutive access requests to the same memory location. The memory cycle time is another parameter that characterizes the overall speed of the memory. The speed is big when the cycle time is small. Memory transfer

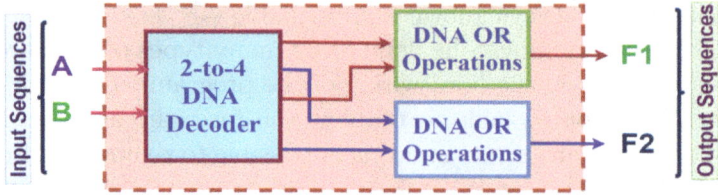

Figure 6.10. Block diagram of 4-to-2 DNA PROM.

rate is the speed of reading or writing data in the given memory, measured in DNA sequences/second.

6.4.5 *Block diagram*

The block diagram of 4-to-2 DNA PROM is shown in Figure 6.10, the unit consists of 4 words of 2 sequences (A = **ACCTAG (1) or TGGATC (0)** and B = **ACCTAG (1) or TGGATC (0)**) each. This implies that there are two output lines (F1 and F2).

6.4.6 *Circuit architecture*

In 4-to-2 DNA PROM circuit architecture (Figure 6.11), there are two DNA NOT and four AND operations that perform DNA 4-to-2 decoder. The output of the decoder then performs two DNA OR operations to produce the desired output functions F1 and F2 of 4-to-2 DNA PROM.

6.4.7 *Working principle*

Truth table of 4-to-2 DNA PROM is given in Table 6.3. The operations in 4-to-2 DNA PROM is explained as follows:

(i) For input combination A, B = **TGGATC**, **TGGATC**, D0 line will be open. So, the value of D0 will be **ACCTAG** and D1 to D3 will be **TGGATC**. For the output of F1, the DNA OR operation will perform between D0 and D2 that produces **ACCTAG**, and for F2, perform DNA OR Operations between D1 and D3 and generates **TGGATC**.

(ii) For input combination A, B = **ACCTAG**, **TGGATC**, D1 line will be open. So, the value of D1 will be **ACCTAG** and D0,

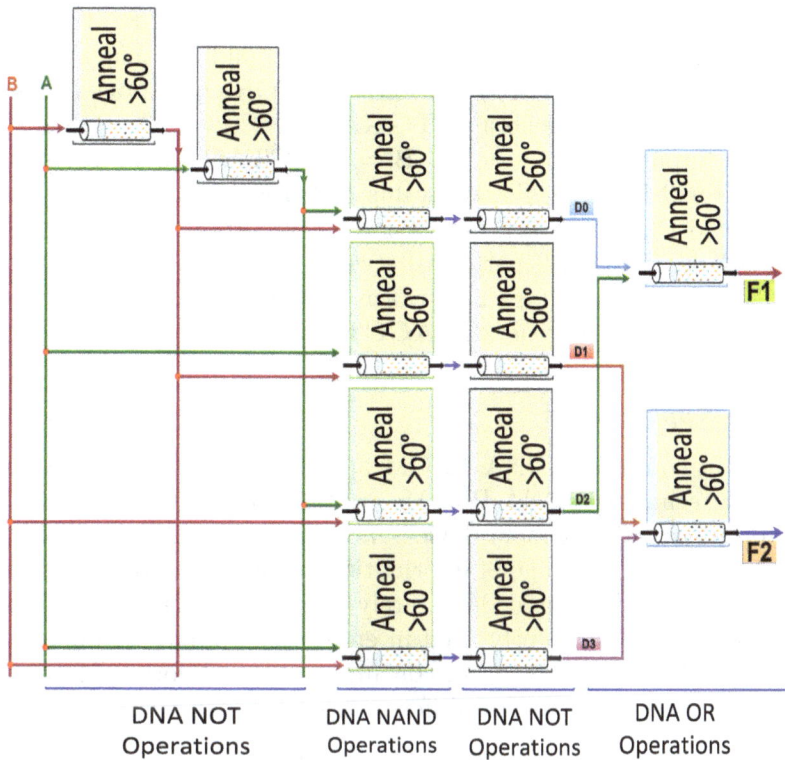

Figure 6.11.   Circuit architecture of 4-to-2 DNA PROM.

Table 6.3.   Truth table of 4-to-2 DNA PROM.

B	A	F1	F2
TGGATC	TGGATC	ACCTAG	TGGATC
TGGATC	ACCTAG	TGGATC	ACCTAG
ACCTAG	TGGATC	ACCTAG	TGGATC
ACCTAG	ACCTAG	TGGATC	ACCTAG

D2, and D3 will be **TGGATC**. For the output of F1, the DNA OR operation will perform between D0 and D2 that produces **TGGATC**, and for F2, perform DNA OR Operations between D1 and D3 and generates **ACCTAG**.

(iii) For input combination A, B = **TGGATC**, **ACCTAG**, D2 line will be open. So, the value of D2 will be **ACCTAG** and D0,

D1, and D3 will be **TGGATC**. For the output of F1, the DNA OR operation will perform between D0 and D2 that produces **ACCTAG**, and for F2, perform DNA OR Operations between D1 and D3 and generates **TGGATC**.

(iv) For input combination A, B = **ACCTAG, ACCTAG**, D3 line will be open. So, the value of D3 will be **ACCTAG** and D0 to D2 will be **TGGATC**. For the output of F1, the DNA OR operation will perform between D0 and D2 that produces **TGGATC**, and for F2, perform DNA OR Operations between D1 and D3 and generates **ACCTAG**.

6.4.8 *Applications*

The main concern of DNA PROM application is data storage and memory technology. DNA PROMs are taken to include all semiconductor, nonvolatile memory devices, and they are used in applications where nonvolatile storage of information, data, or program codes are needed, where the stored data rarely or never change. These types of memories are frequently used in embedded systems or microcontrollers and also in many other consumer and automotive electronics products. They have several different applications:

1. Mobile phones for providing user specific selections.
2. Video game consoles.
3. Implantable medical devices.
4. Radio-frequency identification (RFID) tags.

6.5 DNA Cache Memory

DNA cache memory is the hardware in a computing device where the operating system (OS), application program caches, and data in current use are kept so they can be quickly reached by the device's processor. DNA cache memory is volatile. That means data is retained in cache as long as the computer is on, but it is lost when the computer is turned off. When the computer is rebooted, the OS and other files are reloaded into cache, usually from an HDD or SSD. When the cache fills up, the computer processor must go back and forth to unload that data on the hard disk. The two operations that a cache memory can perform are the write and read operations. The write

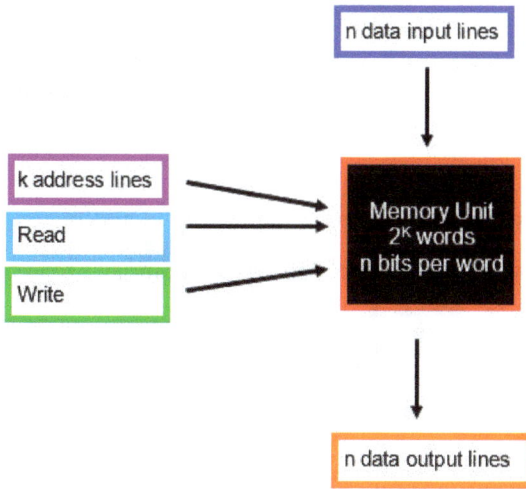

Figure 6.12. 2^k-to-n DNA cache memory.

signal specifies a transfer-in operation and the read signal specifies a transfer-out operation. On accepting one of these control signals. 2^k-to-n cache memory is shown in Figure 6.12. The internal circuits inside the memory provide the desired function. The steps that must be taken to transfer a new word to be stored into memory are as follows:

1. Apply the binary address of the desired word into the address lines.
2. Apply the DNA data sequences that must be stored in memory into the data input lines.
3. Activate the write input.

The memory unit will then take the DNA sequences presently available in the input data lines and store them in the specified by the address lines. The steps that must be taken to transfer a stored word out of memory are as follows:

1. Apply the binary address of the desired word into the address lines.
2. Activate the read input.

The memory unit will then take the DNA sequences from the word that has been selected by the address and apply them to the output data lines. The content of the selected word does not change after reading.

6.5.1 *Advantages*

The disadvantages of DNA-based cache memory are given as follows:

1. Massive storage capacity.
2. Very little energy is required for maintenance.
3. Rapid retrieval of random access memory including algorithmic sorting.
4. Rapid retrieval of large-scale memory.

6.5.2 *Disadvantages*

The major drawback of "DNA-based data storage systems" is the cost involved in writing and reading data on nucleotide sequences. The cost of synthesizing DNA (writing/encoding) is higher than that of sequencing (reading/decoding) in which the processing speed is slow.

6.5.3 *Basic functions*

The main purpose of cache memory in a computer is to read and write any data. Cache memory works with the computer's hard disk. Let us understand this with an example. When opening a Word file, the Word file is stored in the computer's hard disk before it opens, and as soon as the Word file is opened the Word file is stored in the computer's cache memory.

There are many basic and main functions of computer memory cache memory, which are given as follows:

1. **Temporary Storage:** In addition to storing files read from the hard drive, cache memory also stores data that program Caches are actively using but that doesn't need to be saved permanently. By keeping this data in cache memory, program Caches can work with it quickly, improving speed and responsiveness.

2. **Loading Applications:** Loading a software application is also the main function of CACHE. Any software or application opens in the computer using cache itself.
3. **Speed:** DNA cache memory speed is measured in megahertz (MHz), millions of cycles per second so that it can be compared to the processor's clock speed.

6.5.4 *Block diagram*

Figure 6.13 represents the 4-to-1 DNA cache memory, general organization of the block diagram. This DNA cache memory consists of

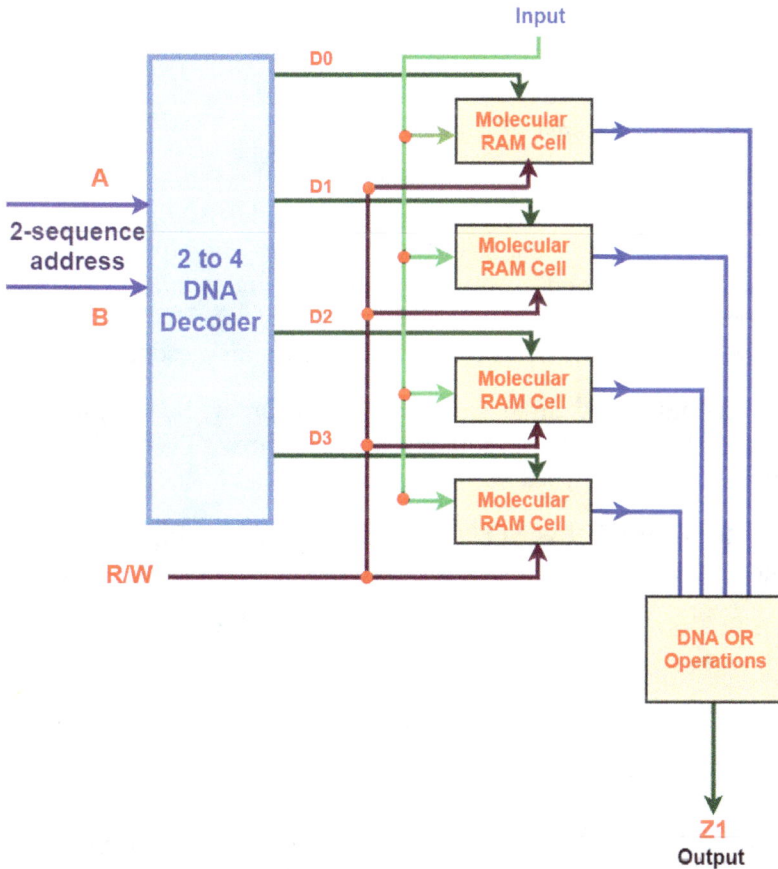

Figure 6.13. Block diagram of 4-to-1 DNA cache memory.

four separate "Words" of memory and each is single sequence wide. The DNA molecular RAM cell has three input sequences and one output sequence.

6.5.5 *Design architecture of basic components*

Cache memory consists of three basic components such as decoder, molecular RAM cells, and OR operations. To execute DNA 4-to-1 Cache memory operation:

1. A 2-to-4 DNA decoder;
2. Molecular RAM cell; and
3. DNA OR operations for corresponding minterms are required.

DNA decoder and DNA OR operations are discussed before.

6.5.5.1 *Circuit design of a molecular cell*

The fundamental design of this molecular cell is based on the D flip-flop (Figure 6.14). To begin with, the molecular RAM cell has three input sequences and a single output sequence. The inputs are labeled "Select," "R/W," and "Input." The output line is labeled "output." To perform the Molecular RAM cell output, two DNA NOT, three DNA AND, and four DNA NAND operations are needed to perform.

Step 1: First draw three input sequences: Input, R/W, and Select. Two possible states for a sequence are the states "**TGGATC**" false, and "**ACCTAG**" true.

Step 2: Draw DNA NOT operation with the Input and R/W sequences.

Step 3: Two input sequences (R/W and select) will go through DNA AND operations.

Step 4: Again, two input sequences (NOT R/W and select) will go through DNA AND operations.

Step 5: The outputs of Step 4 and DNA NOT of sequence input will go to DNA NAND operations. Also, the outputs of Step 4 and sequence input will go to another DNA NAND operations.

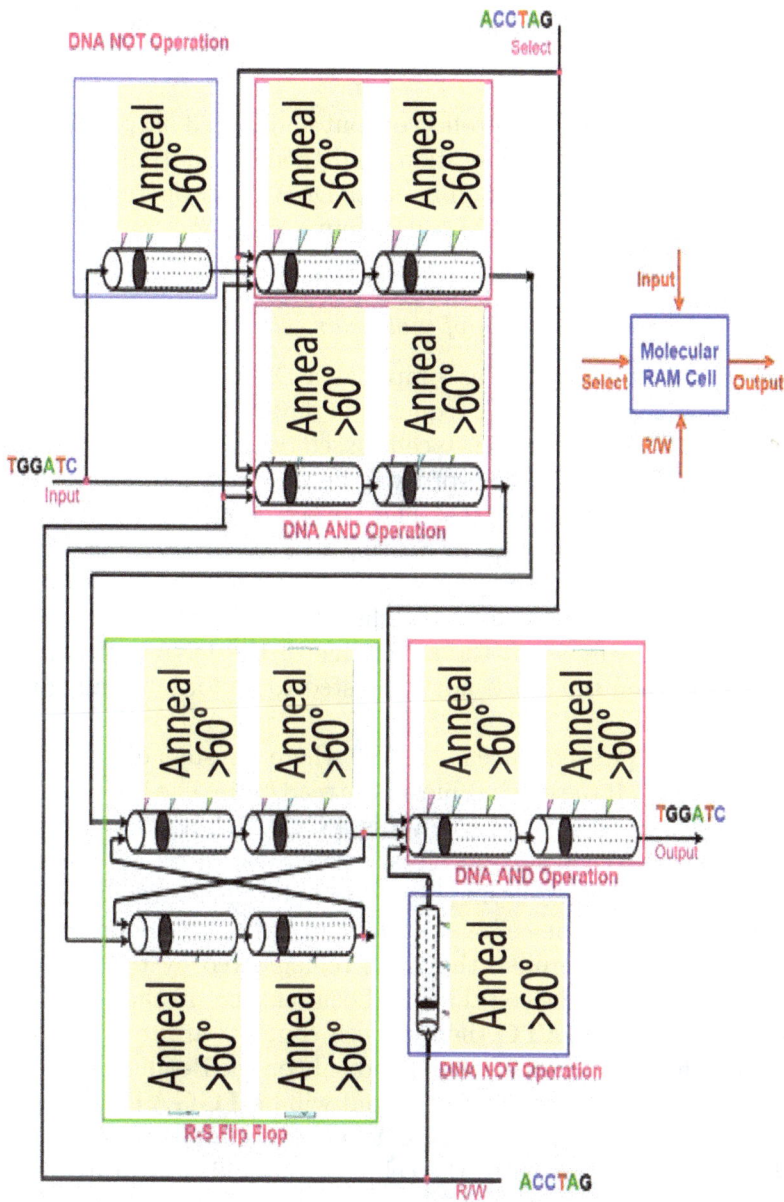

Figure 6.14.   Single molecular DNA RAM cell.

Step 6: The outputs of Step 5 will go to the DNA D flip-flop as input sequence.

Step 7: Finally, DNA D flip-flop output and Step 3 output sequence will go to a DNA AND operation, and then this output with DNA NOT of R/W will go through another DNA AND operation to produce desired molecular RAM cell output sequence.

6.5.5.2 *Working procedure of molecular cell*

In DNA sequential device as simple as a DNA D flip-flop could be used to remember one DNA sequence of data. To develop a complete memory cell, called a sequence cell, based on the flip-flop. The number of total DNA cells per word will be $m \times n$ where m represents words with n DNA sequences. The "select" input is used to access the cell, either for reading or writing also used to access any one sequence cell when there is more than one sequence cell. When the select line is high or **ACCTAG** then the cell performs the memory operation. But when the select line of the DNA cell is low or **TGGATC** then the cell is not interested to perform a read from or written to.

The next input sequence is "R/W" where a system clock will conduct this input. If the clock value on the read/write line is **TGGATC**, this will signify "read" and when it is **ACCTAG**, it will perform the "write" phase. When such a cell is selected and in "read" mode, the current value of its underlying flip-flop will be transferred to the cell's output line. When the cell is selected and in "write" mode, an input data signal will determine the value remembered by the DNA flip-flop. Let's consider the cell that has been selected. In that case, if the clock value is **TGGATC** then the cell contents are to be read and this time the output value will depend only on the P-value of the DNA flip-flop. But if P is low, the cell output will be **TGGATC** and if P is high, the cell output will be **ACCTAG**. It occurs because the DNA AND operation is added to the cell's output which has three input-negated R/W, Select, and P; and both "negated R/W" and "Select are currently high **ACCTAG**. Figure 6.15 shows the four-sequence cells to perform M0 to M3 for further minterms operation.

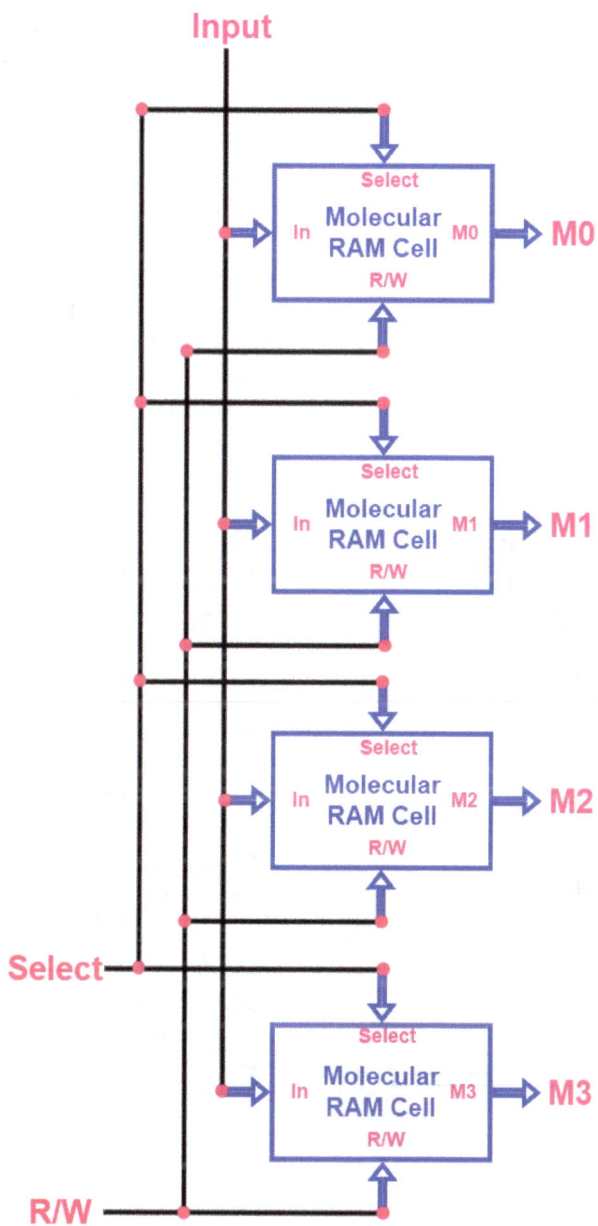

Figure 6.15. Molecular RAM cells.

To perform the 4-to-1 DNA cache memory, four selection lines are needed to design four DNA sequence cells. The output of 2-to-4 DNA decoder with four output sequences which performs as selection input sequence of sequence cell.

6.5.6 *Working principle*

Figure 6.16 represents the implementation of 4-to-1 sequence cache memory. This DNA cache memory consists of four separate "Words" of memory and each is 1 sequence wide. Four words of memory need two address lines. A and B are the two-sequence address lines input that goes through a 2-to-4 decoder that selects one of the four words.

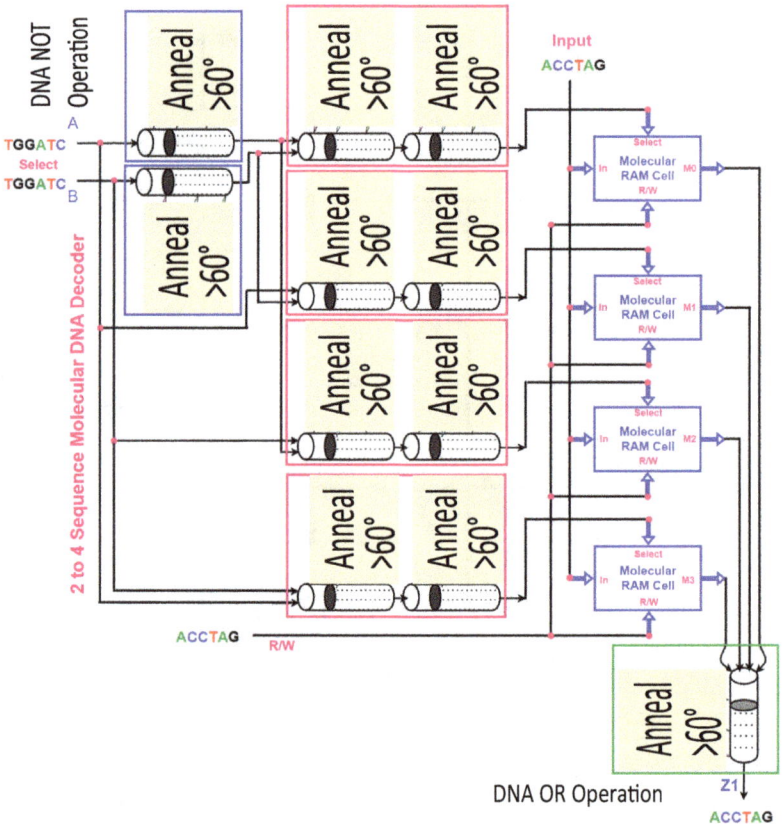

Figure 6.16.   4-to-1 DNA cache memory.

The memory-enabled input enables the decoder. If the memory enables is **TGGATC**, all output of the decoder will be **TGGATC** and in that case, none of the memory addresses will be selected. But when the memory enables is **ACCTAG** then one of the four words is selected. The word is selected by the value in the two address lines. When a word has been selected, the read/write input determines the operation. During the read operation, the four sequences of the selected word pass to the DNA OR operations to the output Z1 terminals. But during the write operation, the data which is available in the input lines are transferred into the four DNA cells of the selected word. The DNA cache memory cells that are not selected are become disabled and their previous sequence never changes. But when the memory enable input that passes into the decoder is equal to **TGGATC**, none of the words are selected, and then all DNA cells remain unchanged regardless of the value of the read/write input.

6.5.7 *Applications*

Most modern operating systems employ a method of extending cache memory capacity, known as "virtual memory." A portion of the computer's hard drive is set aside for a paging file or a scratch partition, and the combination of physical cache memory and the paging file form the system's total memory, for example, if a computer has 2 GB of cache memory and a 1 GB page file, the operating system has 3 GB total memory available to it. When the system runs low on physical memory, it can "swap" portions of Cache memory to the paging file to make room for new data, as well as to read previously swapped information back into cache memory. Excessive use of this mechanism results in thrashing and generally hampers overall system performance, mainly because hard drives are far slower than cache memory.

6.6 Summary

This chapter has presented the details of memory devices in DNA computing where the block diagram, circuit diagram, advantages and disadvantages of DNA memory devices have been shown. DNA computing introduces an approach for generating information that can be stored and retrieved reliably within such a DNA sequence. DNA memory architecture has the capability of long-life storage, low-cost fabrication, and higher memory density. Researchers at UC Davis, the University of Washington, and Emory University have developed a memory technology that applies DNA bases to encode information directly. The researchers have demonstrated the capability to create DNA-based ROM that is programmable and can interface seamlessly with current electronic devices. The technology applies the self-assembly and electrical conductance properties of DNA to create crosswire (X-wire) nanostructures that simulate the "ones and zeroes" that currently form the basis for the electronic storage of digital information. With the development of DNA computing, these memory devices will be available in coming future.

Bibliography

Zamshed I. Chowdhury, Masoud Zabihi, S. Karen Khatamifard, Zhengyang Zhao, Salonik Resch, Meisam Razaviyayn, Jian-Ping Wang, Sachin S. Sapatnekar, and Ulya R. Karpuzcu. A DNA read alignment accelerator based on computational RAM. *IEEE Journal on Exploratory Solid-State Computational Devices and Circuits*, 6(1): 80–88, 2020.

Ivan G. Ivanov, Pencho V. Venkov, and G. George. Isolation OP DNA PROM yeast by chromatography on hydroxyapatite. *Preparative Biochemistry*, 5(3): 219–228, 1975.

Naoto Takahashi, Atsushi Kameda, Masahito Yamamoto, and Azuma Ohuchi. Aqueous computing with DNA hairpin-based RAM. In *International Workshop on DNA-Based Computers*, pp. 355–364. Springer, Italy, 2004.

Zheng-Qing Zhang and Muhammad Ishaque. Evaluation of methods for isolation of DNA from slowly and rapidly growing mycobacteria. *Evaluation*, 65(4): 469–476, 1997.

Chapter 7

DNA Programmable Logic Devices

7.1 Introduction

The four-character genetic alphabets (A [adenine], G [guanine], C [cytosine], and T [thymine]) are used in DNA computing instead of the binary alphabet (1 and 0) utilized by standard computers. Because small DNA molecules of any arbitrary sequence may be manufactured to order, this is possible. The instructions are carried out by laboratory operations on the molecules (such as sorting them by length or chopping strands containing a certain subsequence) and the result is defined as some property of the final set of molecules (such as the presence or absence of a specific sequence).

DNA computing is of twofold: one is theoretical and the other is practical. Starting from observing the structure and dynamics of DNA, the theoretical research begins to propose formal models of DNA computers for performing theoretical operations. The practical side of DNA computing has progressed at a much slower rate, mainly due to the fact that the laboratory work is very time-consuming and includes several constraints.

In this part for programming logical devices, a generic term for an integrated circuit that can be programmed in the laboratory to perform complex tasks. Programmable logic devices in DNA computing are the main concern of this chapter. DNA PLA, DNA PAL, DNA FPGA, and DNA CPLD are the topics to be discussed here.

7.2　DNA Programmable Logic Array

A DNA programmable logic array is a type of DNA PLD, which has both a programmable DNA AND array and a programmable DNA OR array. A DNA PLA is a fixed architecture logic device with programmable DNA AND operations followed by programmable DNA OR operations. PLA is basically a type of programmable DNA logic device used to build a reconfigurable digital circuit. DNA PLDs have an undefined function at the time of manufacturing, but they are programmed before being made into use.

7.2.1　*Block diagram*

DNA PLA is a programmable logic device that has both programmable DNA AND array and programmable DNA OR array. Hence, it is the most flexible PLD. The block diagram of DNA PLA is shown in Figure 7.1. Here, the inputs of DNA AND operation are programmable. That means each DNA AND operation has both normal and complemented inputs of variables. So, based on the requirement, inputs can be programmed.

Here, the inputs of DNA OR operations are also programmable. So, any number can be programmed of required product terms, since all the outputs of DNA AND operations are applied as inputs to each DNA OR operation. Therefore, the outputs of DNA PAL will be in the form of the sum of products form.

7.2.2　*Circuit architecture*

The circuit architecture of DNA PLA is explained in this section. Truth table of DNA PLA for functions F1, F2, and F3 is shown in Table 7.1.

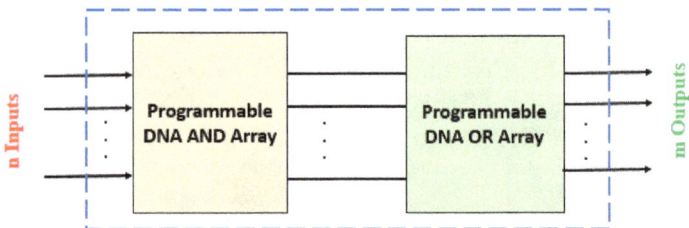

Figure 7.1.　Block diagram of DNA PLA.

Table 7.1.   Truth table of DNA PLA for functions F1, F2, F3.

A	B	C	F1	F2	F3
TGGATC	TGGATC	TGGATC	TGGATC	ACCTAG	TGGATC
TGGATC	TGGATC	ACCTAG	ACCTAG	TGGATC	TGGATC
TGGATC	ACCTAG	TGGATC	TGGATC	ACCTAG	ACCTAG
TGGATC	ACCTAG	ACCTAG	ACCTAG	TGGATC	ACCTAG
ACCTAG	TGGATC	TGGATC	TGGATC	ACCTAG	ACCTAG
ACCTAG	TGGATC	ACCTAG	TGGATC	ACCTAG	ACCTAG
ACCTAG	ACCTAG	TGGATC	ACCTAG	TGGATC	TGGATC
ACCTAG	ACCTAG	ACCTAG	ACCTAG	TGGATC	TGGATC

Let, the F1, F2, and F3 functions as follows:

$$F1\ (A,\ B,\ C) = m\ (1,\ 3,\ 6,\ 7)$$
$$F2\ (A,\ B,\ C) = m\ (0,\ 2,\ 4,\ 5)$$
$$F3\ (A,\ B,\ C) = m\ (2,\ 3,\ 4,\ 5).$$

Express 0 and 1 using DNA molecule:

ACCTAG = TRUE = 1
TGGATC = FALSE= 0

K-map to reduce the functions.

A\BC	00	01	11	10
0	0	1	1	0
1	0	0	1	1

F1 = AB + A$'$C

A\BC	00	01	11	10
0	1	0	0	1
1	1	1	0	0

F2 = A$'$C$'$ + AB$'$

A\BC	00	01	11	10
0	0	0	1	1
1	1	1	0	0

F3 = A$'$B + AB$'$.

Consider the implementation of the following **Boolean functions** using DNA PLA.

$$F1 = AB + A'C$$
$$F2 = A'C' + AB'$$
$$F3 = A'B + AB'.$$

The given three functions are in sum of products form. The number of product terms present in the given Boolean functions F1, F2, and F3 are two. A.B's products are used in both functions F2 and F3. So, five programmable DNA AND operations and three programmable DNA OR operations are required for producing those three functions. The corresponding DNA **PLA** is shown in Figure 7.2.

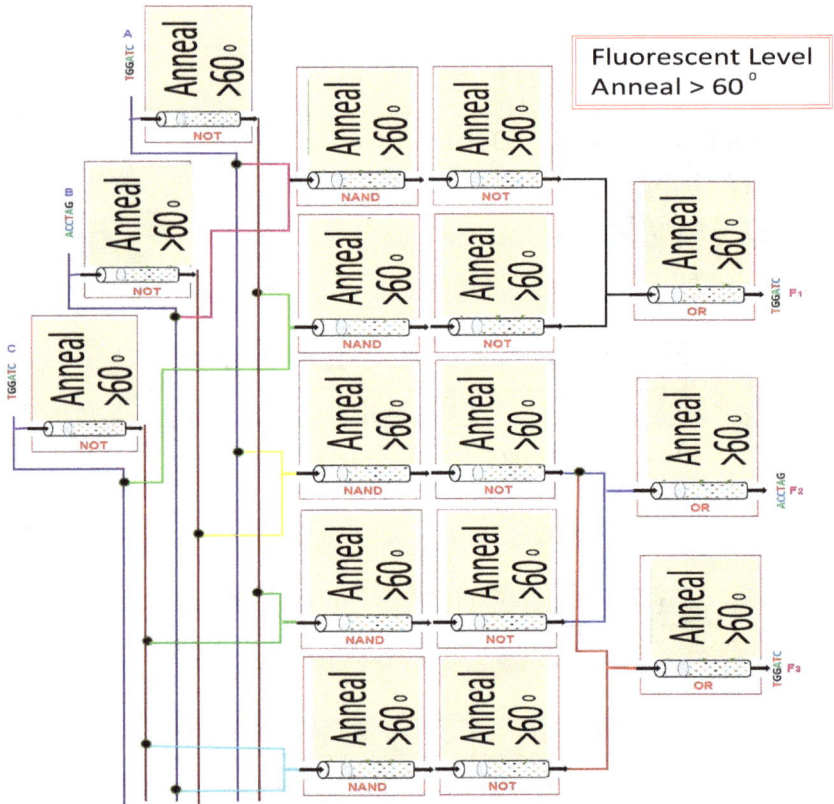

Figure 7.2.   DNA PLA for functions F1, F2, and F3.

Consider the realization of the Boolean expression F1 = AB + A'C, F2 = A'C' + AB', and F3 = A'B + AB' using DNA programmable logic array.

For the given problem, there are three inputs (A, B, C) and three outputs (F1, F2, F3). The complement of three inputs is obtained through DNA NOT operations. Thus, the realization has six input lines (input with its complement).

The given expression has six product terms and so the fuses are placed in the corresponding literals to obtain the product terms.

7.2.3   *Working principle*

According to the truth table (Table 7.1) of DNA PLA, it is necessary to do the following operations to perform the desired output sequence.

1. For input molecular sequences A, B, C = TGGATC, TGGATC, TGGATC function F2 produce ACCTAG.
2. For input molecular sequences A, B, C = TGGATC, TGGATC, ACCTAG function F1 produce ACCTAG.
3. For input molecular sequences A, B, C = TGGATC, ACCTAG, TGGATC function F2 and F3 produce ACCTAG.
4. For input molecular sequences A, B, C = TGGATC, ACCTAG, ACCTAG function F1 and F3 produce ACCTAG.
5. For input molecular sequences A, B, C = ACCTAG, TGGATC, TGGATC function F2 and F3 produce ACCTAG.
6. For input molecular sequences A, B, C = ACCTAG, TGGATC, ACCTAG function F2 and F3 produce ACCTAG.
7. For input molecular sequences A, B, C = ACCTAG, ACCTAG, TGGATC function F1 produce ACCTAG.
8. For input molecular sequences A, B, C = ACCTAG, ACCTAG, ACCTAG function F1 produce ACCTAG.

7.2.4   *Applications*

DNA PLA is used for the implementation of various combinational circuits using DNA AND operation and DNA OR operation. In DNA PLA, all the minterms are not realized but only required minterms are implemented. As DNA PLA has a programmable

DNA AND operation array and programmable DNA OR operation array, it provides more flexibility but the disadvantage is, it is not easy to use.

Applications of DNA PLA are as follows:

1. PLA is used to provide control over the datapath.
2. PLA is used as a counter.
3. PLA is used as a decoder.
4. PLA is used as a BUS interface in programmed I/O.

7.3   DNA Programmable Array Logic

DNA programmable array logic (DNA PAL) is a logic device, which has a programmable DNA AND array and fixed DNA OR array. It is used to realize a logic function. In this DNA PLD, only DNA AND operations are programmable and hence it is easier to work with DPAL. The product terms can be programmed through the fuse link. It means the user can decide the connection between the inputs and the DNA AND operations. If a particular input line is to be connected to the DNA AND operation, then the fuse link must be placed at the interconnection. The DNA AND operation outputs are then fed as an input to the fixed DNA OR operation. Depending upon the required function, the output line of the DNA AND operation is connected to the corresponding input of the DNA OR operation.

7.3.1   *Block diagram*

DNA PAL is a form of programmable logic device (PLD) that may be used to implement a certain logical function. A DNA AND operation array is followed by a DNA OR operation array in DNA PALs. It should be emphasized, however, that only the DNA AND operation array is programmable, whereas the DNA OR operation array has fixed logic. Because the inputs are supplied to the DNA AND operations through fuses that operate as programmable connections, this is the case. When compared to PLAs, DNA PALs have a less flexible programming structure due to their programmable AND and fixed-OR construction (PLAs). The block diagram of **DNA PAL** is shown in Figure 7.3.

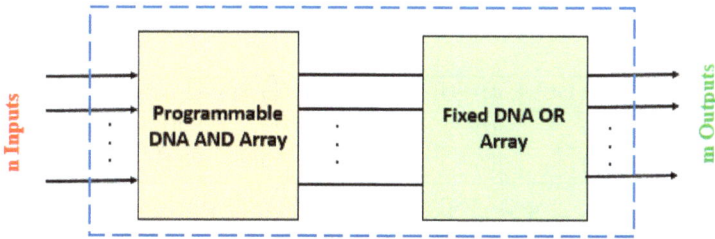

Figure 7.3.   Block diagram of DNA PAL.

Table 7.2.   Truth table of DNA PAL for functions F1 and F2.

A	B	C	F1	F2
TGGATC	TGGATC	TGGATC	TGGATC	ACCTAG
TGGATC	TGGATC	ACCTAG	TGGATC	TGGATC
TGGATC	ACCTAG	TGGATC	TGGATC	ACCTAG
TGGATC	ACCTAG	ACCTAG	ACCTAG	TGGATC
ACCTAG	TGGATC	TGGATC	TGGATC	ACCTAG
ACCTAG	TGGATC	ACCTAG	ACCTAG	ACCTAG
ACCTAG	ACCTAG	TGGATC	ACCTAG	TGGATC
ACCTAG	ACCTAG	ACCTAG	ACCTAG	TGGATC

Here, the inputs of DNA AND operations are programmable. That means each DNA AND operation has both normal and complemented inputs of variables.

7.3.2   *Circuit architecture*

Consider the DNA PAL for the following functions and the truth table of DNA PAL for functions F1 and F2 is given in Table 7.2.

$$F1\ (A,\ B,\ C) = m\ (3,\ 5,\ 6,\ 7)$$
$$F2\ (A,\ B,\ C) = m\ (0,\ 2,\ 4,\ 5).$$

Express 0 & 1 using DNA molecule:

ACCTAG = TRUE = 1
TGGATC = FALSE = 0.

K-map to reduce the functions:

A\BC	$\lvert0\rangle$ $\lvert0\rangle$	$\lvert0\rangle$ $\lvert1\rangle$	$\lvert1\rangle$ $\lvert1\rangle$	$\lvert1\rangle$ $\lvert0\rangle$
$\lvert0\rangle$	$\lvert0\rangle$	$\lvert0\rangle$	$\lvert1\rangle$	$\lvert0\rangle$
$\lvert1\rangle$	$\lvert0\rangle$	$\lvert1\rangle$	$\lvert1\rangle$	$\lvert1\rangle$

$$F1 = AB + BC + AC$$

A\BC	00	01	11	10
0	1	0	0	1
1	1	1	0	0

$$F2 = A'C' + AB'$$

Consider the following **Boolean functions** using DNA PAL:

$$F1 = AB + BC + AC; \text{ and}$$
$$F2 = A'C' + AB'.$$

The given two functions are in sum of products form. The number of product terms present in the given Boolean functions F1 and F2 are three and two, respectively.

Six programmable DNA AND operations and two fixed DNA OR operations are required for producing those two functions. But, it is required to perform an extra two DNA OR operations as three product terms for each function are required to do OR operation. The corresponding DNA **PAL** is shown in Figure 7.4.

Consider the realization of the Boolean expression F1 = AB + BC + AC and F2 = A'C' + AB' using a PLA.

For the given problem, there are three inputs (A, B, C) and two outputs (F1 and F2). The complement of three inputs is obtained through DNA NOT operation. Thus the realization has six input lines (input with its complement). The given first expression has three product terms and the second expression has two product terms. But as the OR operations are fixed the fuses are placed in the corresponding literals to obtain the product terms.

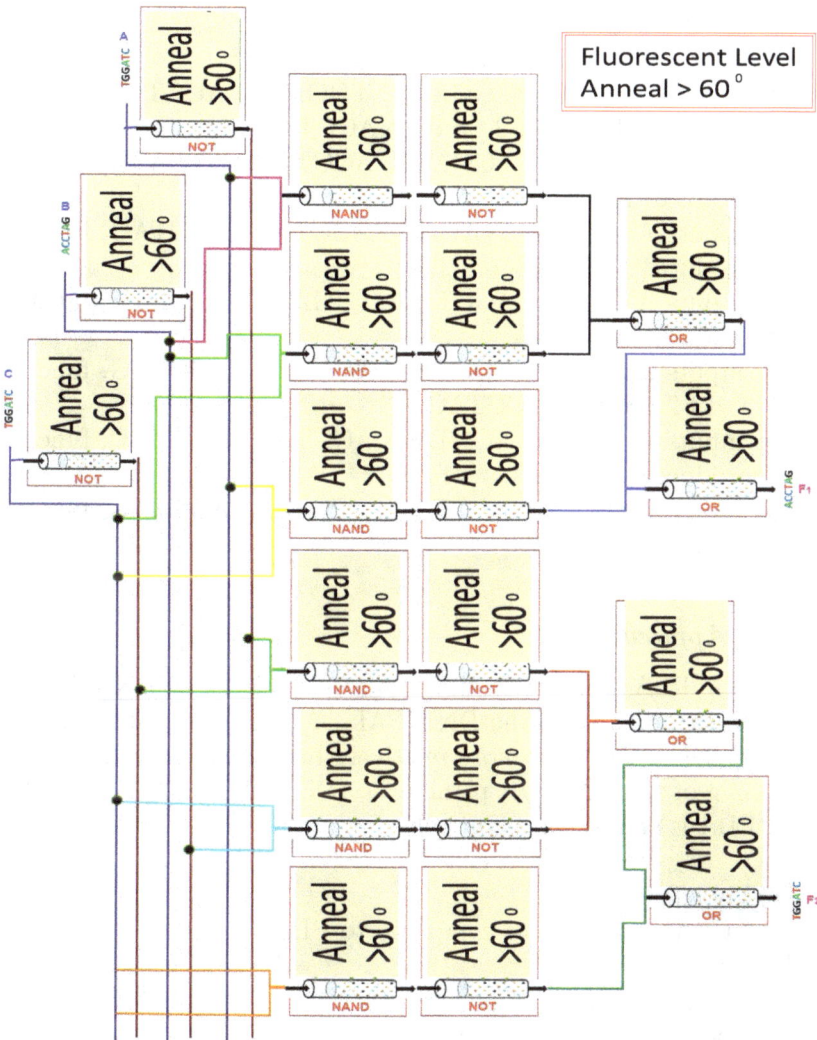

Figure 7.4. DNA PAL for functions F1 and F2.

7.3.3 *Working principle*

According to the truth table (Table 7.2) of DNA PAL, it is necessary to do the following operations to perform the desired output sequence.

1. For inputs A, B, C = TGGATC , TGGATC , TGGATC, function F2 produces ACCTAG.
2. For inputs A, B, C= TGGATC, TGGATC, ACCTAG, none of the functions F1 and F2 produce ACCTAG.
3. For inputs A, B, C = TGGATC, ACCTAG, TGGATC, function F2 produces ACCTAG.
4. For inputs A, B, C = TGGATC, ACCTAG, ACCTAG, function F1 produces ACCTAG.
5. For inputs A, B, C = ACCTAG, TGGATC, TGGATC, function F2 produces ACCTAG.
6. For inputs A, B, C = ACCTAG, TGGATC, ACCTAG, function F1 and F2 produce ACCTAG.
7. For inputs A, B, C = ACCTAG, ACCTAG, TGGATC, function F1 produces ACCTAG.
8. For inputs A, B, C = ACCTAG, ACCTAG, ACCTAG, function F1 produces ACCTAG.

7.3.4 *Applications*

The DNA PLA is more flexible than a DNA PAL. DNA PLA is costlier as compared to the DNA PAL. A number of functions provided by DNA PLA are more relatively because it enables the programming of the DNA OR plane also. DNA PAL works faster while DNA PLA is slower comparatively.

1. Highly efficient.
2. Low production cost as compared to PLA.
3. Highly secure.
4. High reliability.
5. Low power is required for working.
6. More flexible to design.

7.4   DNA Field Programmable Gate Arrays

Traditional silicon-based field programmable gate arrays (FPGAs) is susceptible to security attacks as a consequence of its static design. The reality is, once a static circuit is obtained by an attacker, it is a matter of time before one can reverse engineer its configuration. To circumvent such tampering, circuits must be dynamic by nature.

With this vision in mind, biological methodologies have been developed to mimic existing silicon-based technologies in data manipulation. Within the digital world, data manipulation encompasses data generation, storage, retrieval, and processing. A DNA-based design enables circuitry to be based on biochemical and environmental stimuli. DNA computing introduces a methodology by which one could generate information that could reliably be stored and retrieved within a DNA sequence.

Modern DNA sequencing technologies have greatly reduced their physical dimensions while still being able to process large numbers of molecules. The high information density of DNA molecules and massive parallelism involved in the DNA reactions make DNA computing a powerful tool. It has been proved by many research accomplishments that any procedure that can be programmed in a silicon computer can be realized as a DNA computing procedure. DNA search algorithm is applied to this Boolean equation for solving routing alternatives utilizing the properties of DNA computation. The simulated results are satisfactory and give the indication of the applicability of DNA computing for solving the routing problem.

There are many classical algorithms for finding routing in FPGA. But using DNA computing the routes efficiently can solve and fast. The run time complexity of DNA algorithms is much less than other classical algorithms which are used for solving routing in DNA FPGA. The high information density of DNA molecules and massive parallelism involved in the DNA reactions make DNA computing a powerful tool.

The design of an FPGA using DNA computing also provides some properties such as

1. Dense data storage.
2. Massively parallel computation.
3. Extraordinary energy efficiency.

7.4.1 Block diagram

A basic DNA FPGA architecture consists of thousands of fundamental elements called DNA configurable logic blocks (DNA CLBs) surrounded by a system of programmable interconnects, called a fabric that routes signals between DNA CLBs. DNA FPGA CLB is shown in Figure 7.5.

CLB

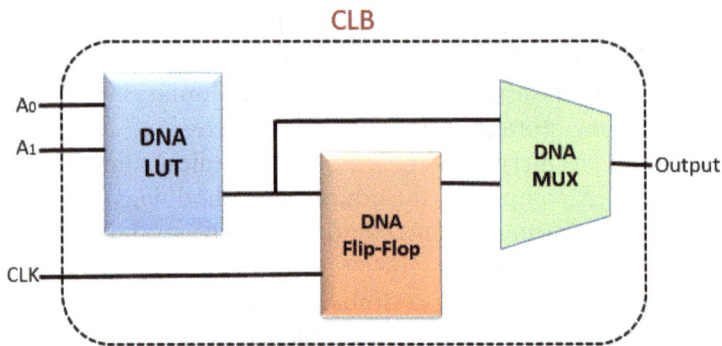

Figure 7.5. FPGA configurable logic block.

A DNA CLB is the basic repeating logic block on a DNA FPGA. There are hundreds of similar logic blocks available onto the DNA FPGA connected via routing resources. The purpose of these logic blocks is to implement combinational and sequential logic.

There are three essential DNA CLB components:

1. DNA flip-flops.
2. DNA look-up tables (LUTs).
3. DNA multiplexers.

7.4.2 *Circuit architecture*

A DNA FPGA logic block is designed by connecting DNA flip-flops, DNA LUTs, and DNA multiplexers. Here a simple DNA FPGA logic block is designed by using a DNA D flip-flop, a DNA LUT, and a DNA multiplexer. A simple 2-input DNA LUT consists of one DNA AND, one DNA NOT, and one DNA OR operations. The output of the DNA LUT will go through the DNA D flip-flop and DNA multiplexer as input. The DNA D flip-flop is a DNA sequential circuit that consists of four DNA NAND operations and one DNA NOT operation. The output of the DNA D flip-flop will go through the DNA multiplexer as input. A 2-to-1 DNA MUX is designed using two DNA AND operations, one NOT operation, and one OR operation. The DNA multiplexer generates the desired output for the DNA FPGA logic. The circuit architecture of a DNA FPGA is shown in Figure 7.6.

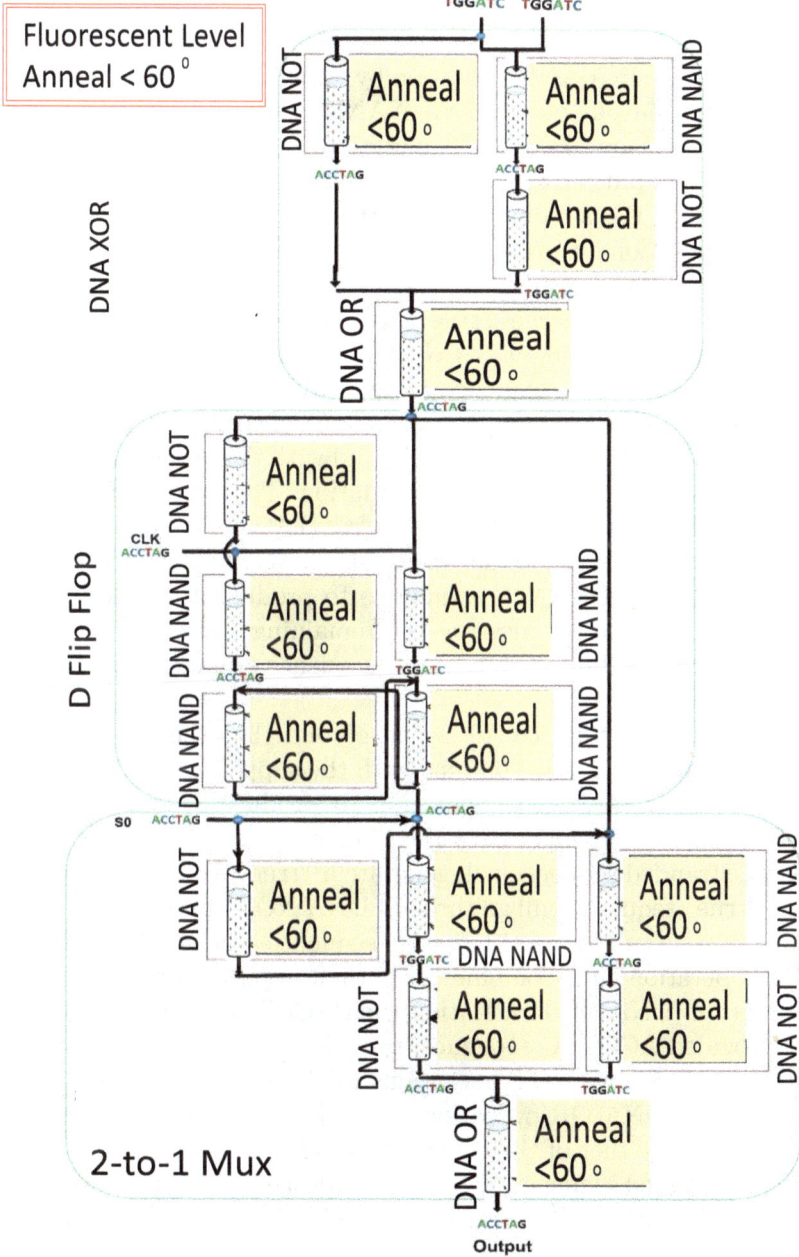

Figure 7.6. Circuit architecture of a DNA FPGA.

7.4.3 *Working principle*

Two inputs A0 and A1 first go through the DNA LUTs. The DNA LUT uses one DNA AND operation by applying the DNA NOT operation to the output of the DNA NAND operation, one DNA NOT operation, and one DNA OR operation. In the DNA LUT, if the both of the input sequences are "false" (TGGATC), then one will combine with the supplied "true" ACCTAG sequence in the DNA NAND operation to produce a double-stranded molecule. DNAase will destroy the remaining input sequence and the double-stranded sequence will generate a "true" ACCTAG sequence. This "true" sequence will go through DNA NOT operation and result in a "false" evaluation. A1 "false" sequence will go through DNA NOT operation and combine with the supplied "true" ACCTAG sequence in DNA NOT operation then ACCTAG will bind with the provided ACCTAG sequence, representing a "true" evaluation. The output of these two DNA NOT operations (TGGATC, ACCTAG) will go through the DNA OR operation, then the "true" ACCTAG sequence will combine with the "false" TGGATC sequences to produce a double-stranded sequence. DNAase will destroy the remaining "false" sequence and this operation will result in a "true" sequence evaluation in the DNA LUT output.

If one input sequence A0 is "false" and the other A1 is "true," then the "false" one will combine with the supplied "true" ACCTAG sequence in DNA NAND operation to produce a double-stranded molecule. DNAase will destroy the remaining input sequence and the double-stranded sequence will generate a "true" ACCTAG sequence. This "true" sequence will go through DNA NOT operation and result in a "false" evaluation. A1 "true" sequence will go through DNA NOT operation and combine with the supplied "true" ACCTAG sequence in DNA NOT operation then ACCTAG will not bind with the provided ACCTAG sequence, provided a "false" evaluation. The output of these two NOT operations (TGGATC, TGGATC) will go through the DNA OR operation, then the "false" sequence will not combine with either of the "false" TGGATC sequences to produce a double-stranded sequence. DNAase will destroy the "false" sequences and this operation will result in a "false" sequence evaluation in the DNA LUT output.

If one input sequence A0 is "true" and the other A1 is "false," then the "false" one will combine with the supplied "true" ACCTAG sequence in DNA NAND operation to produce a double-stranded molecule. DNAase will destroy the remaining input sequence and the double-stranded sequence will generate a "true" ACCTAG sequence. This "true" sequence will go through NOT operation and result in a "false" evaluation. A1 "false" sequence will go through DNA NOT operation and combine with the supplied "true" ACCTAG sequence in DNA NOT operation then ACCTAG will bind with the provided ACCTAG sequence, representing a "true" evaluation. The output of these two DNA NOT operations (TGGATC, ACCTAG) will go through the DNA OR operation, then the "true" ACCTAG sequence will combine with the "false" TGGATC sequences to produce a double-stranded sequence. DNAase will destroy the remaining "false" sequence and this operation will result in a "true" sequence evaluation in the DNA LUT output.

Finally, if both of the input sequences are "true" (ACCTAG), then none will combine with the supplied "true" ACCTAG sequence in DNA NAND operation to produce a double-stranded molecule. DNAase will destroy all sequences and generate a "false" sequence. This "false" sequence will go through DNA NOT operation and result in a "true" evaluation. A1 "true" sequence will go thought DNA NOT operation and combine with the supplied "true" ACCTAG sequence in DNA NOT operation then ACCTAG will not bind with the provided ACCTAG sequence, provided a "false" evaluation. The output of these two DNA NOT operations (ACCTAG, TGGATC) will go through the DNA OR operations, then the "false" sequence will not combine with either of the "false" TGGATC sequences to produce a double-stranded sequence. DNAase will destroy the "false" sequences and this operation will result in a "true" sequence evaluation in the DNA LUT output.

Two-input, one is the output of DNA LUT and another is clock input which will go through the DNA D flip-flop. The DNA D flip-flop uses one DNA NOT operation and four DNA NAND operations. The DNA D flip-flop transfers the DNA LUT output if the CLK input sequence is "true" (ACCTAG). If the CLK input is "false," one of the inputs to each of the last two DNA NAND operations will be "true," thus the output of the DNA D flip-flop remains unchanged regardless of the values of the DNA LUT output.

Two DNA AND operations, one DNA NOT operation and one DNA OR operation are used in 2-to-1 DNA MUX. S_0' is created by applying the DNA NOT operation for input S_0. The DNA NOT operation and DNA LUT output will go through as inputs for the first DNA AND operation. The DNA AND operation is created by applying the DNA NOT operation to the output of the DNA NAND operation. If both of the input sequences are "false" TGGATC sequences, the DNA NOT operation will generate a "true" or ACCTAG sequence for S_0. This "true" sequence and other "false" sequences from the DNA LUT will go through DNA AND operation and result in a "false" evaluation sequence. If one input sequence (S_0) is "false" and the other is "true," the DNA NOT operation will generate a "true" or ACCTAG sequence for S_0. This "true" sequence and other "true" sequences from the DNA LUT will go through DNA AND operation and result in a "true" sequence evaluation. If one input sequence (S_0) is "true" and the other is "false," NOT operation will generate a "false" TGGATC" sequence for S_0. This "false" sequence and other "false" sequence from the DNA LUT will go through DNA AND operation and result in a "true" evaluation sequence. Finally, if the both input sequences are "true" or ACCTAG sequences, the DNA NOT operation will generate a "true" or TGGATC sequence for S_0. This "false" sequence and other "true" sequence from the DNA LUT will go through the DNA AND operation and result in a "false" evaluation sequence.

S_0 and DNA D flip-flop (DFF) output will go through as inputs for the second DNA AND operation. The second DNA AND operation is created by applying the DNA NOT operation to the output of the DNA NAND operation. If both of the input S_0 and DFF output sequences are "false" TGGATC sequences, DNA NAND operation will generate a "true" ACCTAG sequence. This "true" sequence will go through DNA NOT operation and result in a "false" evaluation. If one input sequence is "false" and the other is "true," DNA NAND operation will generate a "true" ACCTAG sequence. This "true" sequence will go through DNA NOT operation and result in a "false" evaluation. Finally, if both input sequences are "true" ACCTAG sequences, DNA NAND operation will generate a "false" sequence. This "false" sequence will go through DNA NOT operation and result in a "true" evaluation.

The output two of DNA AND operations will go through the DNA OR operation to generate the DNA FPGA logical block output. If both of the DNA AND operation output sequences are "true" or ACCTAG sequences, then one sequence will combine with the supplied "false" or TGGATC sequence to produce a double-stranded sequence. DNAase will destroy the remaining input sequence and the double-stranded sequence will result in a "true" evaluation. If one output sequence is "false" and the other is "true," then the "true" ACCTAG sequence will combine with either of the "false" TGGATC sequences to produce a double-stranded sequence. DNAase will destroy the remaining "false" sequence and this operation will result in a "true" evaluation. If both DNA AND operation output sequences are "false" TGGATC sequences, then neither will combine with the supplied "false" sequence. DNAase will destroy all sequences in the mixture, resulting in a "false" evaluation of the DNA FPGA logic block.

7.4.4 *Applications*

Applications of DNA FPGA are given as follows:

1. DNA-based FPGA can perform billions of operations simultaneously.
2. DNA-based FPGA computing can provide huge memory in a small space.
3. Developing DNA-based FPGA logic circuits might be a new direction of Nanoscale computing with great possibility of implementation in the biomedical field.
4. An exemplifying advantage of DNA-based FPGA, besides their fundamental significance, pertains to circuit security, providing for obfuscation and tamper-proofing.
5. The benefits of a DNA-based FPGA logic circuit are not limited to the reduction in the number of operations based on the additional representation of an additional output state; it also enables circuits to be compressed based on inputs.

7.5   DNA Complex Programmable Logic Device

The acronym of the DNA CPLD is "complex programmable logic devices," it is one kind of integrated circuit that application designers design to implement digital hardware like mobile phones. These can handle knowingly higher designs than DNA simple programmable logic devices (SPLDs) but offer less logic than DNA FPGAs. DNA CPLDs include numerous logic blocks; each of the blocks includes 8–16 macrocells. Because every logic block executes a specific function, all of the macrocells in a logic block are fully connected. Depending upon the use, these blocks may or may not be connected to one another. Most DNA CPLDs have macrocells with a sum of logic functions, a DNA D FF and a DNA MUX. Depending on the chip, the DNA logic function supports from 4 to 16 product terms with inclusive fan-in. DNA CPLDs also differ in terms of DNA shift registers and DNA logic operations. Due to this reason, DNA CPLDs with a huge number of DNA logic operations may be used instead of DNA FPGAs. Another DNA CPLD specification signifies the number of product terms that a macrocell can accomplish. Product terms are the product of digital signals that execute a specific logic function. DNA CPLDs are available in several IC package forms and logic families. DNA CPLDs also differ in terms of supply voltage, operating current, standby current, and power dissipation.

7.5.1   *Block diagram*

A complex programmable logic device comprises a group of DNA programmable functional blocks (FBs). Most DNA CPLDs have macrocells with a sum of logic functions, a DNA D FF and a DNA MUX. DNA CPLD functional block is shown in Figure 7.7.

A DNA FB is the basic repeating logic block on a DNA CPLD. The function blocks have programmable interconnections. The DNA programmable FB looks like the array of DNA logic operations, where an array of DNA AND operations can be programmed and DNA OR operations are stable. A switch matrix is used for function blocks to function block interconnections. Further, the switch matrix in a DNA CPLD may or may not be fully connected. The complexity of a typical DNA PAL device is around a few hundred DNA logic

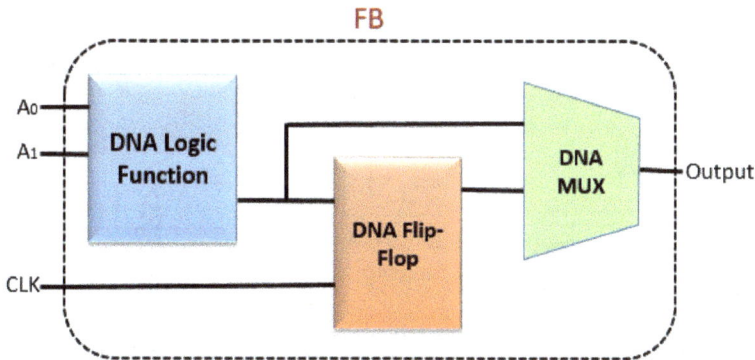

Figure 7.7. DNA CPLD functional block.

operations whereas the complexity of DNA CPLD is around tens of thousands of DNA logic operations. The DNA CPLDs have predictable timing characteristics hence are suitable for critical control applications and other applications where a high-performance level is required. Further, due to low power consumption and low cost, DNA CPLDs are mostly used for battery-operated portable applications such as mobile phones, digital assistants, etc.

There are three essential DNA FBs components:

1. DNA flip-flops;
2. DNA logic function; and
3. DNA multiplexers.

7.5.2 *Circuit architecture*

DNA CPLD FB is designed by connecting DNA flip-flops, DNA logic function and DNA multiplexers. Here a simple DNA CPLD FB is designed by using a DNA D FF, a DNA logic function and a multiplexer. A simple three input logic function ($F = AB + BC + CA$) consists of three DNA AND and two DNA OR operations. Output of the logic function will DNA XOR with zero. The output of the DNA XOR will go through the DNA D FF and DNA multiplexer as input. The DNA D FF is a DNA sequential circuit that consists of four DNA NAND operations and one DNA NOT operation. The output of the D FF will go through the multiplexer as input. A 2-to-1 DNA MUX

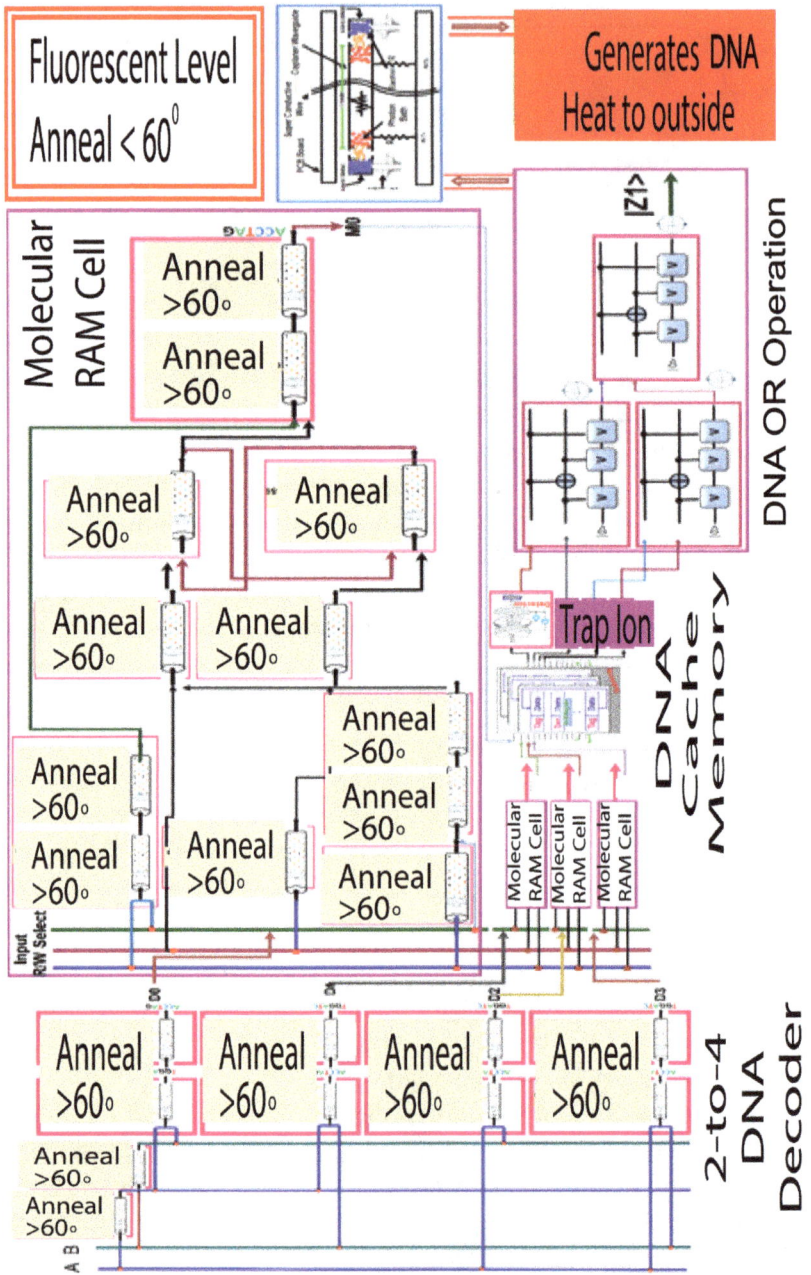

Figure 7.8.　Circuit architecture of DNA CPLD.

is designed using two DNA AND operations, one DNA NOT operation, and one DNA OR operation. The DNA multiplexer generates the desired output for the DNA CPLD FB. The Circuit architecture of DNA CPLD is shown in Figure 7.8.

7.5.3 *Working principle*

Three inputs A, B, and C first go through the logic block. The logic function is implemented using three DNA AND operation and two DNA OR operation. Outputs of the logic function will go through the DNA XOR (DNA XOR with TGGATC) operation and produce the output. If all the three input sequences are "false" (TGGATC), then all the inputs will go through three DNA NAND operations and combine with the supplied "true" ACCTAG sequence in DNA NAND operation to produce a double-stranded molecule. DNAase will destroy the remaining input sequence and the double-stranded sequence will generate a "true" or ACCTAG sequence. These "true" sequences will go through DNA NOT operation and result in a "false" evaluation. The output of these three DNA NOT operations will go through the DNA OR operation. DNase will destroy the "false" sequence and the operation will result in a "false" evaluation in the logic function output.

Two inputs, one is the output of DNA XOR and another is clock input will go through the DNA DFF. The DNA D FF uses one DNA NOT operation and four DNA NAND operations. The DNA D FF transfer the DNA XOR output, if the CLK input sequence is ACCTAG. If the CLK input is TGGATC, one of the inputs to each of the last two DNA NAND operations will be ACCTAG, thus the output of the DNA D FF remains unchanged regardless of the values of the DNA XOR output.

Two DNA AND operations, one DNA NOT operation and one DNA OR operation are used in 2-to-1 DNA MUX. S_0' is created by applying the DNA NOT operation for input S_0. The DNA NOT operation and DNA logic function output will go through as inputs for the first DNA AND operation. The DNA AND operation is created by applying the DNA NOT operation to the output of the DNA NAND operation. If both of the input sequences are "false" or TGGATC sequences, DNA NOT operation will generate a "true" or ACCTAG sequence for S_0. This "true" sequence and other "false" sequences (DNA logic function) will go through DNA AND operation and result in a "false" evaluation. If one input sequence (S_0) is "false" and the other is "true," DNA NOT operation will generate a "true" or ACCTAG sequence for S_0. This "true" sequence and other "true" sequences (DNA logic function) will go through DNA AND operation and result in a "true" evaluation. If one input sequence (S_0) is "true" and the other is "false," DNA NOT operation will generate a "false" or TGGATC" sequence for S_0. This "false" sequence and other "false" sequence (logic function) will go through DNA AND operation and result in a "true" evaluation. Finally, if both input sequences are "true" or ACCTAG sequences, DNA NOT operation will generate a "true" or TGGATC sequence for S_0. This "false" sequence and other "true" sequence (DNA logic function) will go through DNA AND operation and result in a "false" sequence evaluation.

S_0 and DNA D FF output will go through as inputs for the second DNA AND operation. The second DNA AND operation is created by applying the DNA NOT operation to the output of the DNA NAND operation. If both of the input S_0 and DNA D FF output sequences are "false" or TGGATC sequences, DNA NAND operation will generate a "true" or ACCTAG sequence. This "true" sequence will go through DNA NOT operation and result in a "false" sequence evaluation. If one input sequence is "false" and the other is "true," DNA NAND operation will generate a "true" ACCTAG sequence. This "true" sequence will go through DNA NOT operation and result in a "false" evaluation. Finally, if both input sequences are "true" ACCTAG sequences, DNA NAND operation will generate a "false" sequence. This "false" sequence will go through NOT operation and result in a "true" evaluation.

The output of two AND operations will go through the DNA OR operation to generate the DNA CPLD Function block output. If both of the AND operation output sequences are "true" or ACCTAG sequences, then one sequence will combine with the supplied "false" or TGGATC sequence to produce a double-stranded sequence. DNase will destroy the remaining input sequence and the double-stranded sequence will result in a "true" evaluation. If one output sequence is "false" and the other is "true," then the "true" or ACCTAG sequence will combine with either of the "false" or TGGATC sequences to produce a double-stranded sequence. DNase will destroy the remaining "false" sequence and this operation will result in a "true" evaluation. If both AND operation output sequences are "false" or TGGATC sequences, then neither will combine with the supplied "false" sequence. DNAase will destroy all sequences in the mixture, resulting in a "false" evaluation of the DNA CPLD function block.

7.5.4 *Applications*

DNA CPLDs find their application in many low-to-medium DNA complexity digital control and signal processing circuits. Some of the important applications are listed as follows:

1. DNA CPLDs can be used as bootloaders for DNA FPGAs and other DNA programmable systems.
2. DNA CPLDs are often used as address DNA decoders and DNA custom state machines in DNA systems.
3. Due to their small size and low power consumption, DNA CPLDs are ideal for use in portable handheld digital devices.
4. DNA CPLDs are also used in safety-critical control applications.
5. DNA complex programmable logic devices are ideal for high-performance, critical control applications.
6. DNA CPLD can be used in digital designs to perform the functions of the boot loader.
7. DNA CPLD is used for loading the configuration data of a DNA FPGA from DNA non-volatile memory.
8. Generally, these are used in small design applications like address decoding.
9. DNA CPLDs are frequently used in many applications like in cost-sensitive, battery-operated portable devices due to their low size and usage of low power.

7.6 Summary

PLDs in DNA computing are discussed here. Individual applications, working principles of DNA PLDs mean DNA PLA, DNA PAL, DNA FPGAs, DNA CPLDs are discussed in detail in this chapter. DNA computing needs extra heat to perform their calculations. So, the external source of heat is needed to supply extra heat to the DNA circuit during computations. The next chapter is about the PLDs in DNA computing.

Bibliography

Leonard M. Adleman. Molecular computation of solutions to combinatorial problems. *Science*, 266(5187): 1021–1024, 1994.

George M. Church, Yuan Gao, and Sriram Kosuri. Next-generation digital information storage in DNA. *Science*, 337(6102): 1628, 2012.

Yaniv Erlich and Dina Zielinski. DNA fountain enables a robust and efficient storage architecture. *Science*, 355(6328): 950–954, 2017.

Christy M. Gearheart, Eric C. Rouchka, and Benjamin Arazi. DNA-based dynamic logic circuitry. In *2010 53rd IEEE International Midwest Symposium on Circuits and Systems*, pp. 248–251. IEEE, 2010.

Larry J. Kricka and Paolo Fortina. Analytical ancestry: "Firsts" in fluorescent labeling of nucleosides, nucleotides, and nucleic acids. *Clinical Chemistry*, 55(4): 670–683, 2009.

Rizki Mardian and Kosuke Sekiyama. Ant systems-based DNA circuits. *Bio-NanoScience*, 5(4): 206–216, 2015.

Lee Organick, Siena Dumas Ang, Yuan-Jyue Chen, Randolph Lopez, Sergey Yekhanin, Konstantin Makarychev, Miklos Z. Racz, Govinda Kamath, Parikshit Gopalan, Bichlien Nguyen, *et al.* Random access in large-scale DNA data storage. *Nature Biotechnology*, 36(3): 242–248, 2018.

Ankur Sarker, Hafiz Md Hasan Babu, and Sarker Md Mahbubur Rashid. Design of a DNA-based reversible arithmetic and logic unit. *IET Nanobiotechnology*, 9(4): 226–238, 2015.

Shivashekar Murali Shankar. *Prediction of Protein Coding Regions in DNA Sequences Using the Stockwell Transform*. PhD thesis, Citeseer, 2011.

Chapter 8

DNA Nano Processor

8.1 Introduction

In 1994, Leonard M. Adleman, a professor of computer science and molecular biology at the University of Southern California, created a storm of excitement in the computing world when he built the first deoxyribo nucleic acid (DNA)-based computer capable of solving complex combinatorial problems. For example, his DNA computer solved a directed Hamiltonian path problem, better known as the traveling salesman problem. He solved this problem using DNA and molecular chemistry. His incredible work has led to a new hope of making a DNA computer quickly solving nano processor (NP) combinatorial problems.

Each successful instance may have a different model, algorithms, and biological operations, but the basic idea of simulating DNA computing is the same. There are many approaches for performing DNA computing which all are test tube-based processes. Test tube-based model is a more widespread process of DNA computing that is used to verify whether the algorithm is feasible or not. In this chapter, an integrated scheme based on a test tube approach for performing DNA computing to integrate all the steps of DNA computing into one machine is called computing NP, in which annealing temperatures, DNase enzyme may have predetermined DNA base sequence present or not on the test tube. When the input and base sequence bind, the DNA sequence makes a double-stranded bond. If the arrangement makes a bond with each other, it represents a true result. Otherwise, it means a false result. When the sequences make a bond, then the

DNase enzyme can not break or destroy the double-stranded bond, but if they are not able to make a bond the DNase enzyme destroys both of the sequences, and in that case, the output is false or low.

8.2   Basic Definitions

A DNA NP is a combinational DNA circuit that can execute instructions on behalf of programs made of DNA molecular scale. It performs all kinds of NP combinatorial problems and logical units carry out the instruction of the computer program. The NP controls all types of data flow and instructions. However, The CPU consists of five core components which are as follows:

1. DNA control unit (DCU);
2. DNA register;
3. DNA arithmetic logic unit (DNA ALU);
4. DNA RAM;
5. DNA buses;
6. DNA control unit.

The control unit is the main component of a NP in a DNA computer that directs the operation of the NP. The primary function is to fetch and execute instructions from the main memory.

1. **DNA Register:** Registers are one kind of memory used to accept, store, and transfer data and instructions used immediately by the NP. The primary function is to hold an instruction, a storage address, or any data. The register mainly holding the memory location will be used to calculate the address of the next instruction when the current instruction is completed.
2. **DNA Arithmetic Logic Unit:** In computing, an arithmetic logic unit is a combinational DNA circuit that handles all the calculations the NP may need. It performs all types of arithmetic operations like addition, subtraction, and division.
3. **DNA RAM:** RAM stands for "Random Access Memory," called volatile memory. It is one of the components that contain information as a DNA sequence, and the NP can read or write to those sequences as information depending on whether the READ or WRITE line is signaled. The main goal of RAM is to store and access data on a short-term basis.

4. **DNA Buses:** Buses are high-speed internal connections used to send control signals and data between the NP and other components of the DNA computers. All computers need buses, whether they are DNA computers, quantum computers, supercomputers, or classical computers; they are used for transferring data between processors and other components. That is why The DNA NP used in DNA computers will have been described here, where buses will be used like other computers. There are three types of buses — address bus, control bus, and data bus, briefly described in the fundamental components section.

8.3   Block Diagram of Complete DNA Nano Processor

The basic block diagram of a DNA NP consists of nano-arrays unit, nano control unit, nano RAM, encoding DNA sequences unit, nano ALU. All circuits are made of DNA logic as a DNA NP. The red, blue, and green lines represent the control bus, data bus, and address bus. A complete DNA NP block diagram is given in Figure 8.1.

The working procedure of a 2-molecular DNA NP is like a 2-molecular DNA NP. This is a real DNA NP showing all fundamental components. To accomplish meaningful work, CPUs need to have two inputs-instruction and data. The main function of instructions is to give directions to the CPU on what actions need to be performed on the data. In DNA NPs, instructions are represented using DNA sequences: ACCTAG and TGGATC, which are respectively described as "True" and "False." The CPU's inputs are stored in the memory. According to Figure 8.1, data is moving from RAM to the instructions register through.

The data bus, which is shown uses blue lines. The CPU functions always follow a cycle of fetching instructions from memory. After fetching, it is decoded, and finally, it is executed. The CPU cycle does not work until data is transferred from memory to the instruction register. The unique DNA sequence patterns are extracted by selecting machine language in the instruction register and sent to the decoder. The primary purpose of the DNA decoder is to decode coded information from one format to another format and the second step of the cycle starts to work because of the decoder. The decoder mainly represents which DNA sequence patterns will operate and activates that circuit needed to perform the actual operation. The circuit will

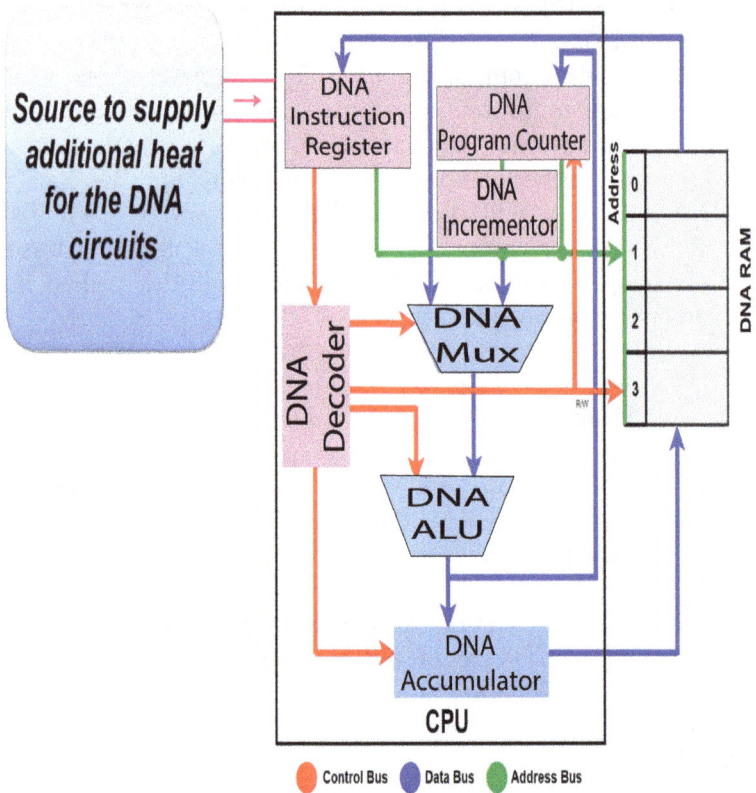

Figure 8.1.   2-Molecular DNA nano processor.

work with the next instruction when the procedure is thoroughly accomplished. The program counter (PC) holds the address of the next instruction to be executed from memory known to the CPU. If an instruction is completed, the PC is incremented by one memory location. This is the whole working procedure of this DNA NP.

8.4   Basic Components of DNA Nano Processor with Design Procedures and Working Principle

A complete authentic DNA NP has been made using the following components shown in Figure 8.1. The features of this CPU are given as follows:

1. DNA RAM;
2. DNA instruction register;
3. DNA program counter;
4. DNA incrementor;
5. DNA decoder;
6. DNA multiplexer;
7. DNA ALU;
8. DNA accumulator.

For transferring data from one component to another, buses are used. There are three types of buses: data bus, address bus, and control bus. The data bus is bidirectional, which carries the data between the processor and other components. Bus width is the most crucial feature of the data bus. It means the width of a data bus refers to the number of DNA sequences of DNA that the bus can carry at a time. For example, this NP is built of 2-molecular DNA sequences, and it can take the 2-molecular sequences simultaneously between the NP and main memory. The main function of the address bus is to carry memory addresses from the NP to other components like primary storage and input/output devices. And the last bus is a control bus that carries the control signals from the NP to other parts. These are also essential components of the CPU that need to be accomplished for meaningful work.

Now all components of the DNA NP will be described in detail.

8.4.1 *Design procedure of DNA RAM*

Two DNA sequences are required as address lines, and each address line has to be connected with a DNA NOT operation for simulating 4-to-2-molecular DNA RAM. These address line combinations will be the input of 2-to-4 DNA decoders which consists of four DNA AND functions, and the DNA decoder must enable input. Four select lines are achieved from the decoder and each select line is attached to each RAM cell. The word calculation of RAM will be 2^k where k is the address line, 2^k is the total words of n date sequences, and the decoder combination will be $k \times 2^k$. This 2-molecular RAM consists of four separate DNA RAM cells, and each cell has inputs such as D0 or D1. Anyone selects lines and read/write inputs. The obtained output from four DNA RAM cells will be the input of a DNA OR operation operation, which produces the final output. This is the whole design procedure of 4-to-2-molecular RAM.

8.4.1.1 *Working principle of DNA RAM*

This is the most crucial CPU component that stores the data as a DNA sequence for a short time, a primary and volatile memory. The circuit of DNA RAM is given in Figure 8.2.

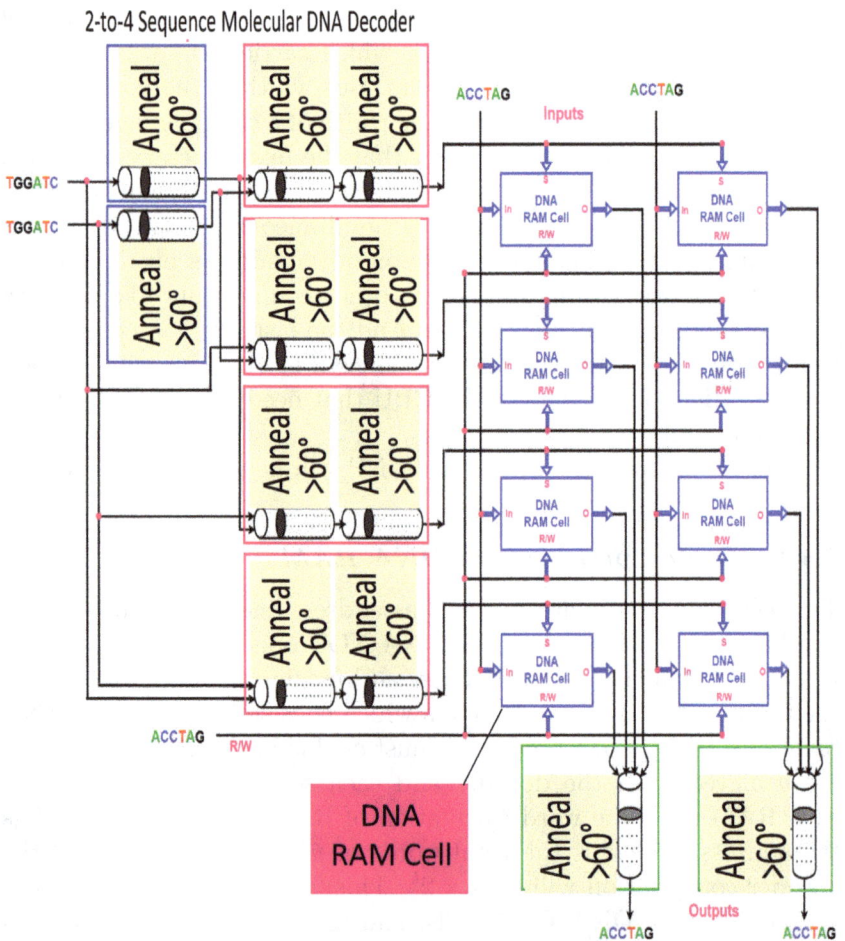

Figure 8.2. 2-Molecular DNA RAM.

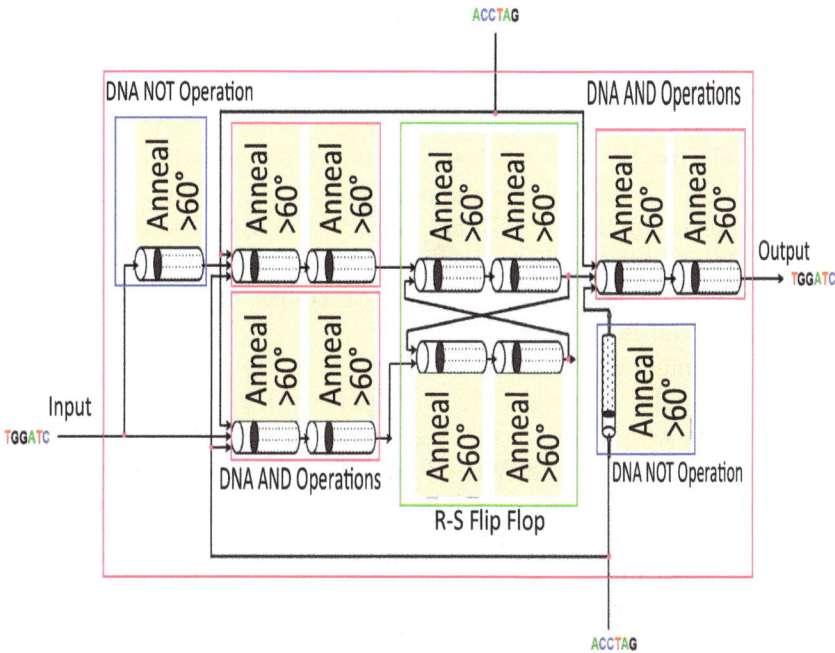

Figure 8.3.   DNA RAM cell.

Figure 8.3 represents the implementation of 4-to-2-molecular sequence DNA RAM. This DNA RAM consists of four separate "Words" of memory, and each is 2-molecular wide. This RAM cell contains three inputs and one output. The complete circuit of a DNA RAM cell is described in Figure 8.3 with a proper explanation. A word consists of two DNA RAM cells arranged in such a way so that both DNA sequences can be accessed simultaneously. Four terms of memory need two address lines. For example, A0 and A1 are the 2-molecular DNA sequence address lines input that goes through a 2-to-4 decoder that selects four words.

The memory-enabled information enables the decoder. If the memory enable is 0, i.e. TGGATC, all outputs of the decoder will be TGGATC, and in that case, none of the memory addresses will be selected. But when the memory enable is one or ACCTAG, one of the four words is selected. The word is selected by the value in the two address lines. The read/write input determines the operation when a word has been selected. The four DNA sequences of the selected word pass to the DNA OR operation to the output ACCTAG and ACCTAG terminals during the read operation. But during the write operation, the data available in the input lines are transferred into the four DNA cells of the selected word. The DNA RAM cells that are not selected become disabled, and their previous DNA sequence never changes. But when the memory-enabled input that passes into the decoder is equal to TGGATC, none of the words is selected, and then all DNA cells remain unchanged regardless of the value of the read/write input. This is the working procedure of DNA RAM. The DNA RAM cell is given in Figure 8.3.

To implement DNA RAM cell DNA R-S flip-flop has been used here. The number of total DNA cells per word will be $m \times n$, where m represents words with n DNA sequences. This cell contains three inputs — "Select," "Read/Write," and "Input," and one output line that is labeled by "Output." The "select" input is used to access either reading or writing. The cell performs the memory operation when the select line is high or ACCTAG. But when the select line of the DNA cell is low or TGGATC, the cell is not interested in completing a read from or written to. The following input is "Read/Write," where a system clock will conduct this input. If the clock value on the read/write line is TGGATC, this will signify "read," and when it is ACCTAG, it will perform the "write" phase. Let's consider the cell that has been selected. In that case, if the clock value is TGGATC, then the cell contents are to be read, and this time the output value will depend only on the Q value of the flip-flop. But if Q is low, the cell output will be TGGATC, and if Q is high, the cell output will be ACCTAG. It occurs because the DNA AND circuit added to the cell's output have three inputs — negated read/write, select, and Q; and both "negated read/write" and "select are currently high."

8.4.2　*Design procedure of DNA instruction register*

The DNA instruction register consists of 16 DNA AND operations shown in Figure 8.4. The instruction DNA sequence is double the CPU DNA sequence. So, as this instruction is a component of a two DNA sequences CPU, instructions will be $2 \wedge 2 = 4$. Therefore, the minimum four instructions can be defined as LOAD A, LOAD B, ADD A B, and OUT. Instruction register is a particular register

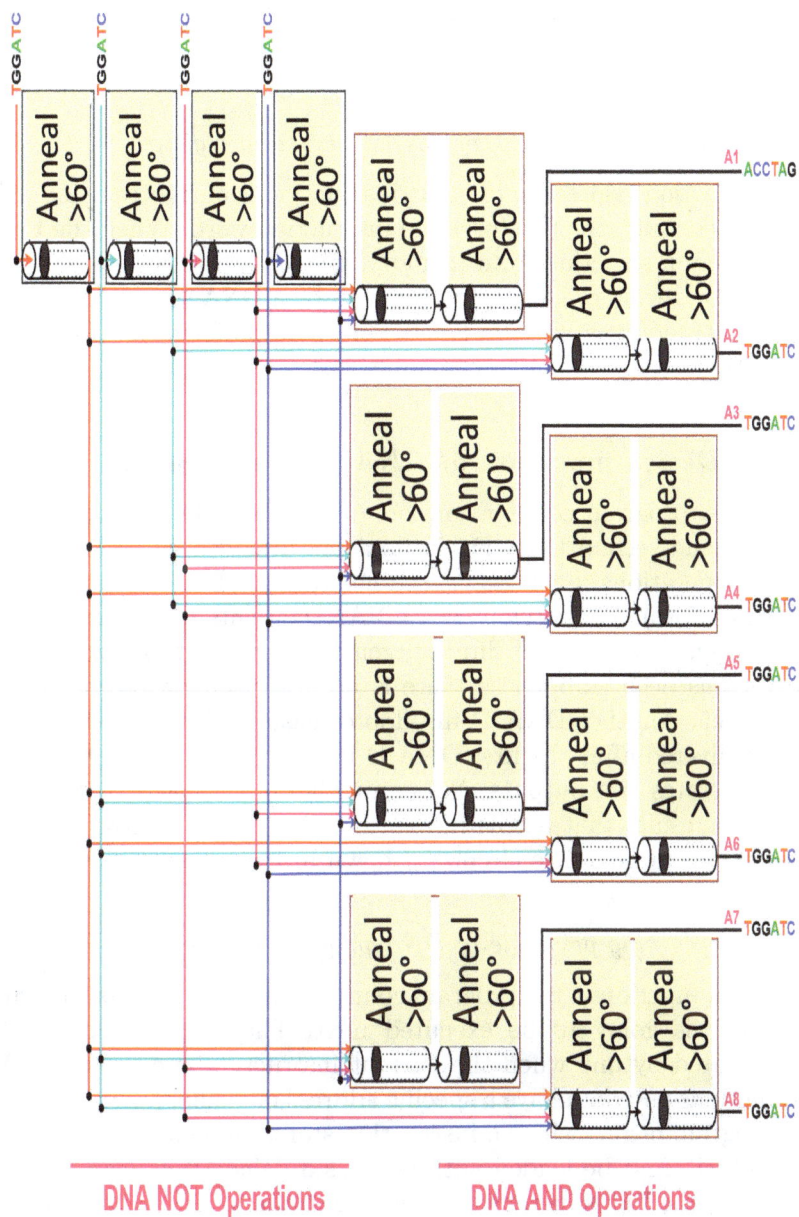

Figure 8.4. DNA instruction register.

mainly used to store the instructions currently being executed by the DNA computer.

8.4.2.1 *Working principle of DNA instruction register*

An instruction register holds that instruction is currently being executed. Generally, the IR stores the instruction word. When the CPU fetches any instruction from memory, it is temporarily stored in the instruction register. The instruction can be a DNA sequence word or code that defines a specific operation to be performed. After that, the CPU decodes the instruction and then executes it.

8.4.3 *Design procedure of DNA program counter*

The DNA program counter consists of two DNA D flip-flops where the first and second DNA D flip-flops are designed using four DNA NAND operations. Each input also has to be in DNA in DNA NOT operation form. The top four DNA NAND circuits represent the first D-flip flop. This DNA D flip-flop generates two outputs where one output has been skipped as there is no need to show, according to Figure 8.5. Another DNA D flip-flop consists of four NAND operations delivered after the first DNA D flip-flop. This DNA D flip-flop also generates two outputs. Notice in Figure 8.5 that it is required only two results that act as inputs. So, the rest of the two outputs have been skipped. The circuit is shown in Figure 8.5.

8.4.3.1 *Working principle of DNA program counter*

The main function of the program counter is to store the next instruction that is going to be executed next. The program counter is incremented by one when the current instruction is completed. All instructions and data have a specific address in memory. For example, if a program begins with an instruction stored in memory location 3, the PC will first be loaded with address 3. When this instruction is executed, the PC is incremented by one to the following address, i.e. 4. The instructions in a program always follow the sequence memory location for storing themselves. In the above example, ACCTAG and TGGATC represent input that goes through to the D-flip flop and gives the desired output.

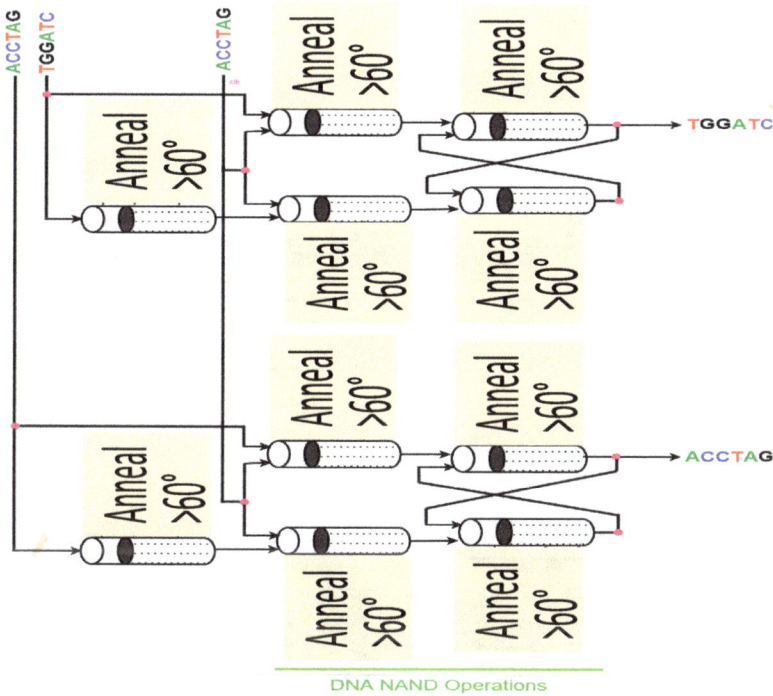

Figure 8.5.   2-Molecular DNA program counter.

8.4.4   *Design procedure of DNA incrementor circuit*

The DNA incrementor circuit is designed only using two DNA half adder circuits shown in Figure 8.6, where each DNA half adder consists of an individual DNA XOR circuit and individual DNA AND circuit. DNA XOR operation is used to compute the sum result, and DNA AND operation is the carry-out result.

8.4.4.1   *Working principle of DNA incrementor circuit*

The PC, known as the instruction pointer, is a special purpose register that holds the address of each instruction and tells the CPU in what order they should be carried out. When an instruction is being executed, one needs to be incremented, repeatedly repeating until it reaches the STOP instruction. That is why a 2-molecular DNA incrementor circuit is used here, where a one DNA sequence is added to with that value stored in a register. Consider the above

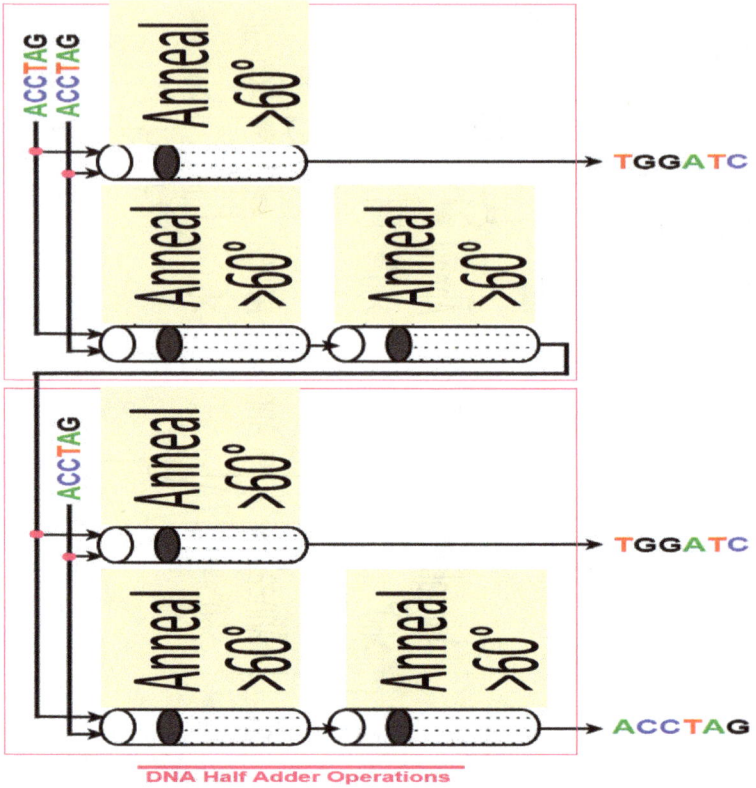

Figure 8.6. 2-Molecular DNA incrementor.

example where two ACCTAG are given as inputs; where the first half adder's DNA XOR operation performs addition and delivers the output TGGATC, and a DNA AND operation produce carry out is the input of the second half adder circuit. During performing the second half adder, the DNA XOR circuit produces TGGATC as output, and DNA AND circuit deliver ACCTAG.

8.4.5 *Design procedure of DNA decoder*

To implement a 2-molecular DNA CPU, a 2-to-4 DNA decoder is used where this decoder consists of four DNA AND operations with one enable input. Two inputs are fed into this decoder, and each input must be a DNA NOT operation form. And this decoder produces

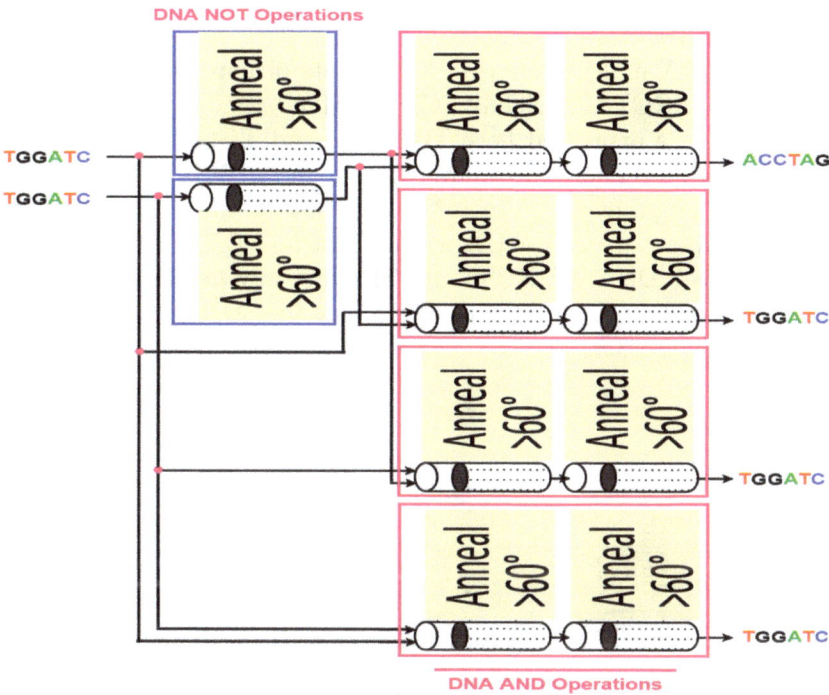

Figure 8.7.   2-Molecular DNA decoder.

four outputs based on given input sequences. The following circuit is 2-molecular DNA decoder in Figure 8.7.

8.4.5.1   *Working principle of DNA decoder*

The DNA decoder is a combinational DNA circuit that may have 2-molecular, 3-molecular, or 4-molecular DNA sequences based on the number of data input lines, so a decoder that contains a set of two or more DNA sequences will be defined as having an n DNA sequences code and will produce 2^n output lines. A decoder generally decodes a DNA sequence. However, one product is achieved from each output, and to gain this product, DNA AND operations are performed. Consider the above example where two inputs are TGGATC. When enable is equal to one, the minterm of two input variables will be TGGATC. But if enable is zero, all the decoder outputs will be similar to zero, i.e. TGGATC.

8.4.6   *Design procedure of DNA multiplexer*

A 4-to-1 DNA multiplexer (MUX) consists of four data input lines defined as ACCTAG, TGGATC, ACCTAG, and TGGATC; two select lines are expressed here TGGATC and TGGATC and a single output line. There are many ways to draw 4-to-1 DNA MUX, but here designing the 4-to-1 MUX by using three 2-to-1 MUX. The calculation of DNA MUX will be $2 \wedge n$-to-1 DNA MUX requires $(2 \wedge n\text{-}1)$2-to-1 DNA MUX. S0 and S1 select any four input lines to connect the output lines. The circuit of 4-to-1 DNA multiplexer is shown in Figure 8.8.

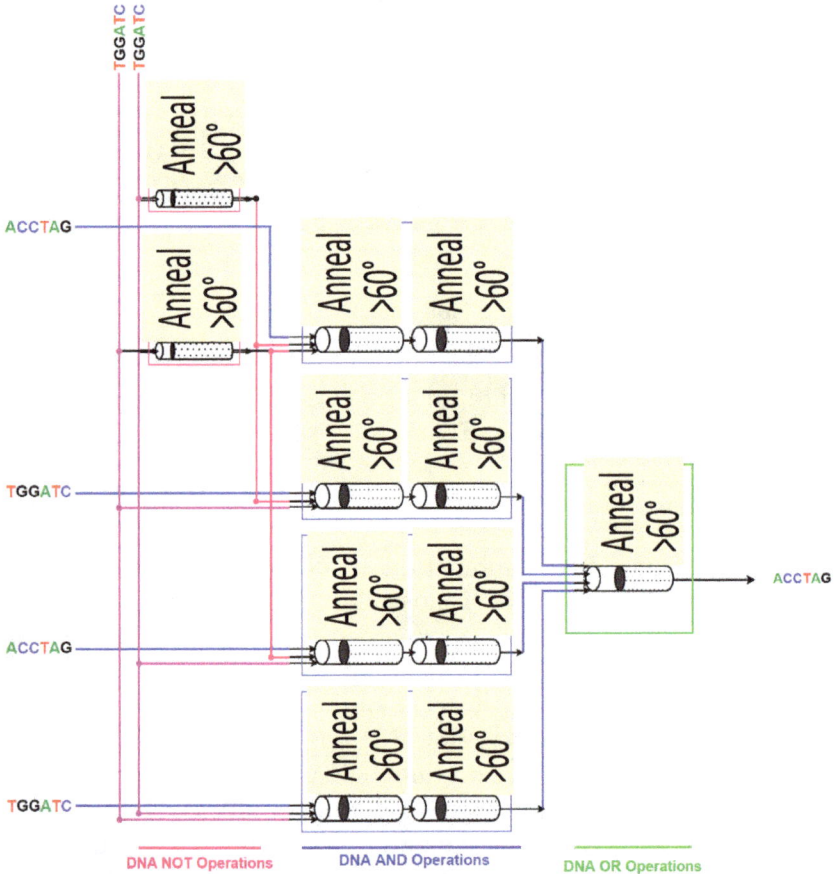

Figure 8.8.   4-to-1 DNA multiplexer.

8.4.6.1 *Working principle of DNA multiplexer*

The DNA MUX has multiple inputs and a single output. Each 2-to-1 DNA MUX consists of two DNA AND operations. The obtained production from two DNA AND operations will be the input of a DNA OR operation. The two outputs are obtained from 2-to-1 DNA MUX, and these two outputs are propagated to the final 2-to-1 DNA MUX which is the final output.

8.4.7 *Design procedure of DNA ALU*

A DNA ALU is the core component of a NP. It is a DNA combinational circuit that performs DNA arithmetic and DNA logic operations. The DNA control unit indicates to DNA ALU what process it needs to achieve on the data and stores the result in an output register. It is required to construct a 2-molecular DNA ALU for a 2-molecular DNA NP. This 2-molecular DNA ALU consists of 16 DNA AND operations selected by 2-to-4 DNA decoder. According to the 2-to-4 DNA decoder, only one logical operation is performed, such as DNA addition, DNA multiplication, DNA subtraction, or DNA division. The following circuit is 2-molecular DNA ALU that is shown in Figure 8.9.

Every logical operation like addition and subtraction is executed with the help of a decoder. Each logical operation is described as follows:

• **DNA Adder:** This DNA adder consisting of two DNA full adders is used to perform addition. The circuit architecture of DNA adder is shown in Figure 8.10.

According to the diagram in Figure 8.10, the first full adder produces sum and the Cout that will be the Cin of the next full adder. And this full adder finally produces a second output and a Cout. The design procedure and working principle of ternary DNA full adder are discussed in the previous chapters.

• **DNA Subtractor:** This subtractor as shown in Figure 8.11 consisting of the DNA full subtractors is used to perform subtraction. A full subtractor is such a combinational DNA circuit that can perform subtraction of two molecules where A and B are control DNA sequences, Bin is the borrow input and output is the difference D,

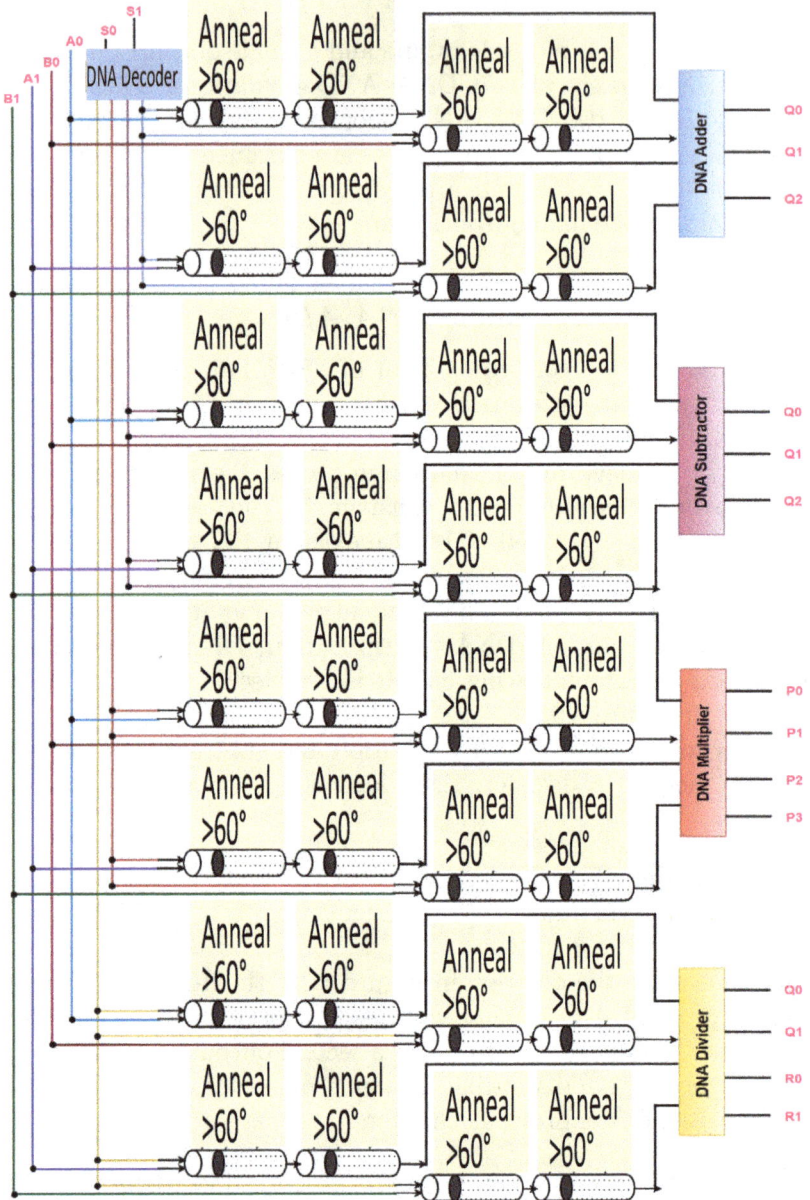

Figure 8.9.   DNA 2-molecular ALU.

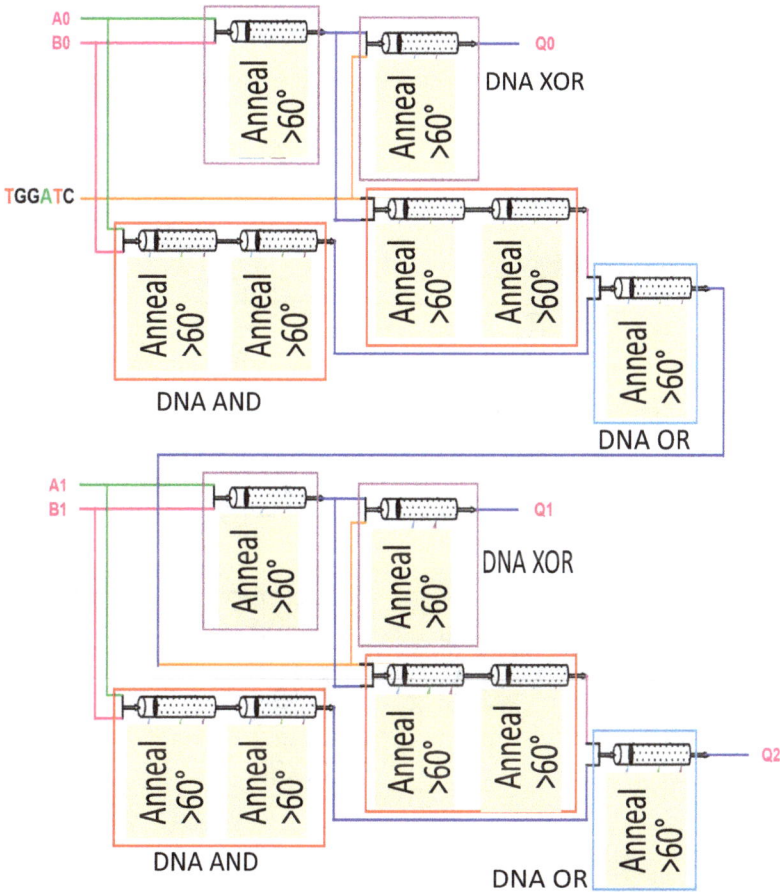

Figure 8.10. DNA adder.

and Bout is the borrow out. Here A, B indicates respectively minuend and subtrahend, and 0 is considered a target DNA sequence
which is constant here. According to Figure 8.11, the first full subtractor produces a difference, and Bout will be the Bin of the next
full subtractor. And this full subtractor finally has a second output
and a Bout.

• **DNA Multiplier:** For performing a 2-molecular DNA multiplier, it is required two DNA sequences ACCTAG and TGGATC
that represent, respectively, 1 and 0. In this circuit, four DNA AND
operations have been implemented with two half adders. Here, the

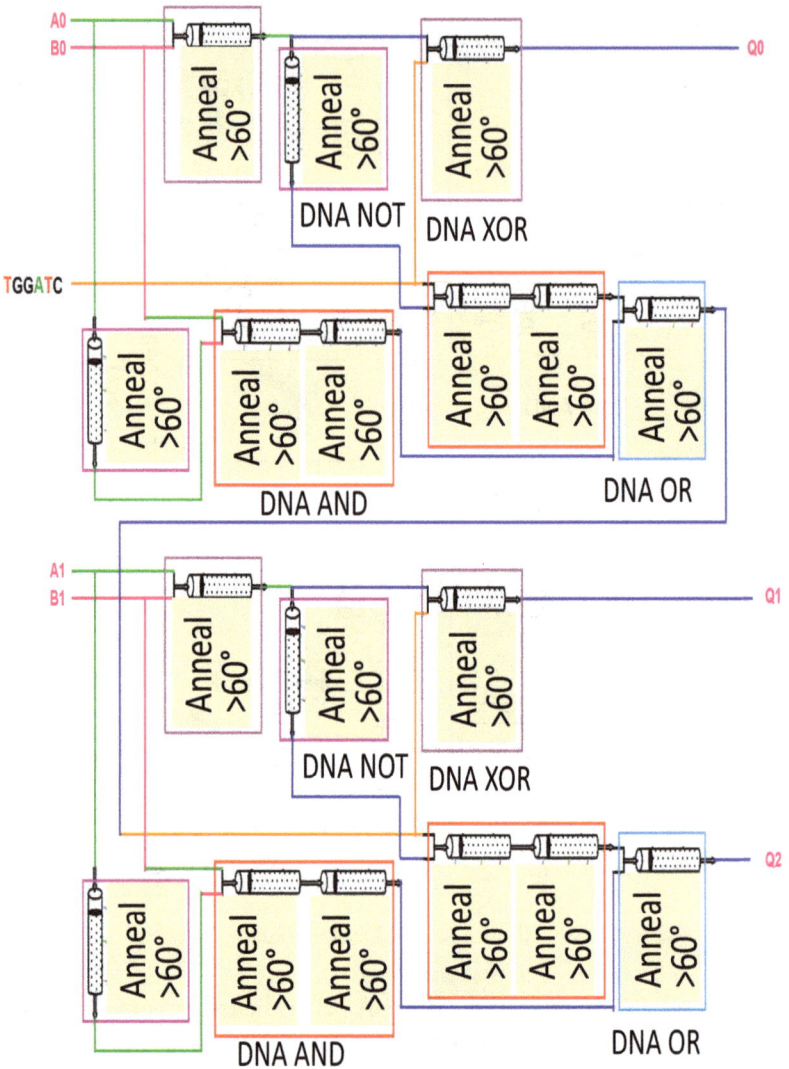

Figure 8.11. DNA subtractor.

DNA AND operations will perform the DNA multiplication and DNA half adders will add the DNA partial product terms or DNA carry. Hence the circuit obtained is given in Figure 8.12.

At first, it is required to perform four DNA AND operations for performing DNA multiplication and DNA half adders will add the

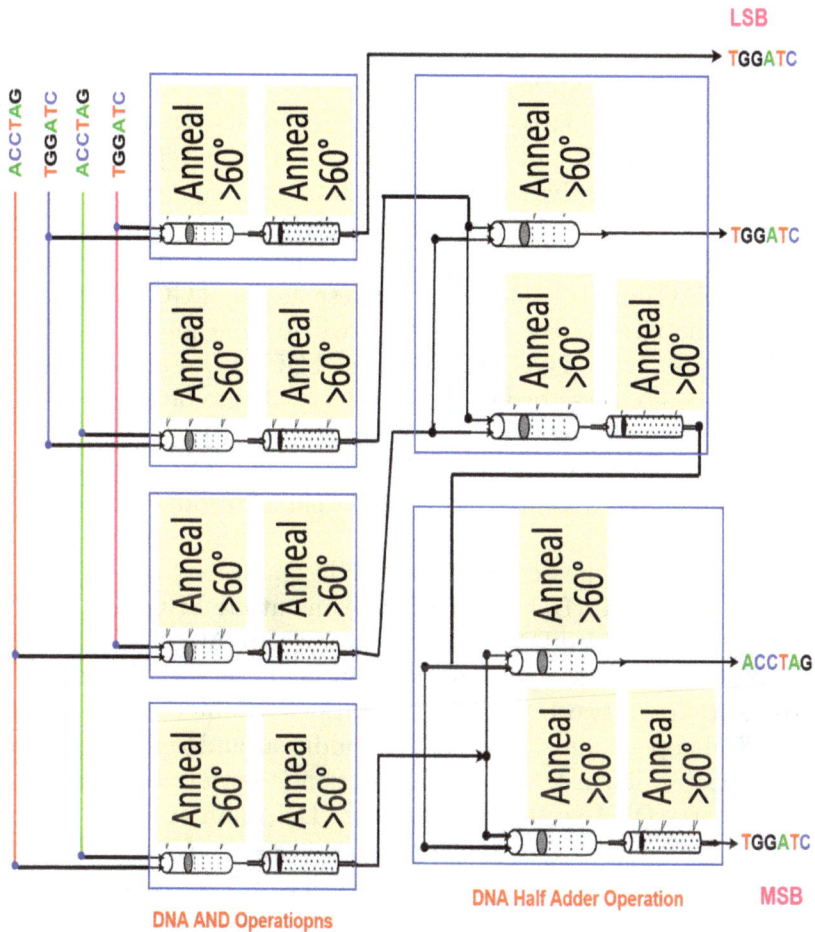

Figure 8.12.   DNA multiplier.

partial product. In this circuit, an input portion the first two DNA sequences of left sides are ACCTAG, TGGATC which are multiplier and second two DNA sequences of right sides are also ACCTAG, TGGATC which are multiplicand. ACCTAG and TGGATC represent respectively 1 and 0. There is a constant DNA base sequence in AND operation tubes which makes either TRUE or FALSE bond with the upcoming input sequence. Their TRUE bond will be double-strand but if they do not make any bond, it will make a single bond which is destroyed by the DNase enzyme. They will make a true (1) bond when ACCTAG and TGGATC coincide. But when two ACCTAG sequences or two TGGATC sequences try to make a bond, then their bonding will be false because the same sequence will never be able to make a double-strand bond. So, in that case, the output will be 0. The TRUE and FALSE bond respectively represents 1 and 0. According to the given circuit, the first DNA sequence of input is B_1 which is represented by ACCTAG, the second DNA sequence of input is B_0 (TGGATC), the third DNA sequence of input represents A_1 (ACCTAG) and the last input represents A_0 (TGGATC).

From the circuit, the first output will be the multiplication of TGGATC (A_0) and TGGATC (B_0). As TGGATC represents 0, the multiplication of these two DNA sequences will also be 0 (TGGATC) that is the first output.

The first partial product of LSB is the first output/product of LSB. Therefore, it's no need to add any partial product. So, output obtained directly.

The inputs of the second DNA AND operations are ACCTAG (A_1) and TGGATC (B_0) which will give an output TGGATC (0) and the inputs of the third DNA AND operations are ACCTAG (B_1) and TGGATC(A_0) which gives output TGGATC (0). The obtained two outputs from the two DNA AND operations will be the input of the DNA XOR operation which perform addition and generate output TGGATC (0). Then DNA AND operation is performed which generates a carry DNA sequence. This is the DNA half adder operation. And finally, the last DNA AND operation gives output ACCTAG (1) that is the multiplication of ACCTAG (B_1) and ACCTAG (A_1). This output and the previous carry DNA sequence from the first half adder will be the input of DNA XOR operation that gives output ACCTAG (1) and the DNA AND operation of a DNA half adder will be the carry DNA sequence. As there is no carry DNA sequence, it simply represents TGGATC (0). Now, for going MSB (Most significant DNA sequence) to LSB (least significant DNA sequence) the output will be, respectively, TGGATC (0), AGGTAC (1), TGGATC (0), TGGATC (0) that is the production of ACCTAG (1), TGGATC (0) multiplicand and ACCTAG (1), TGGATC (0) multiplier. According to the DNA logic, the product of 10 molecules and 10 molecules is 0100 which has been implemented here. In this circuit, the input can be anything. So, it is the whole working procedure of a 2-molecular DNA multiplier.

$$\textbf{A}_1 \textbf{ (ACCTAG)} \quad \textbf{A}_0 \textbf{ (TGGATC)}$$

$$\textbf{B}_1 \textbf{ (ACCTAG)} \qquad \textbf{B}_0 \textbf{ (TGGATC)}$$

$$\textbf{A}_1\textbf{B}_0 \textbf{ (FALSE) } \textbf{A}_0\textbf{B}_0 \textbf{ (FALSE) } _{(Partial\ product)}$$

$$\textbf{B}_1\textbf{A}_1 \textbf{ (TRUE) } \textbf{B}_1\textbf{A}_0 \textbf{ (FALSE) } \times \text{ (Left shift)}$$

$$\textbf{C}_2 \textbf{ (FALSE) } \textbf{B}_1\textbf{A}_1 \textbf{ (TRUE) } \textbf{A}_1\textbf{B}_0 \textbf{ (FALSE) } \textbf{A}_0\textbf{B}_0 \textbf{ (FALSE)}$$

$$+$$

$$\textbf{B}_1\textbf{A}_0 \textbf{ (FALSE)}$$

$$\textbf{P}_3 \textbf{ (FALSE) } \textbf{P}_2 \textbf{ (TRUE) } \textbf{P}_1 \textbf{ (FALSE) } \textbf{P}_0 \textbf{ (FALSE)}_{(Total\ output)}$$

If the multiplication of any two DNA sequences is 0, it will be false that is expressed by F and if production is 1 of any two DNA sequences, it will be true that is represented by 1 in the above multiplication portion.

• **DNA Divider:** The logical operations of the circuit depend on the number of DNA sequences. This circuit has been completed using a 2-molecular divisor and 2-molecular divisible which produce a maximum of four DNA sequences output. Here, DNA sequence means sequence ACCTAG (1) and TGGATC (0). Let, the divisor is ACCTAG (1) and TGGATC (0) and the dividend is also ACCTAG (1) and ACCTAG (1). Now, if ACCTAG (1), ACCTAG (1) is dividend by divisor ACCTAG (1), TGGATC (0), then quotient respectively 0 at first output and 1 at second output which works according to the following circuit. Figure 8.13 shows the circuit architecture of DNA divider operation.

8.4.7.1 *Working principle of DNA ALU*

A DNA ALU is a component of a DNA NP capable of performing logical operations. All information is stored in a DNA computer and manipulated in 0 and 1 where 0 and 1, respectively, TGGATC and

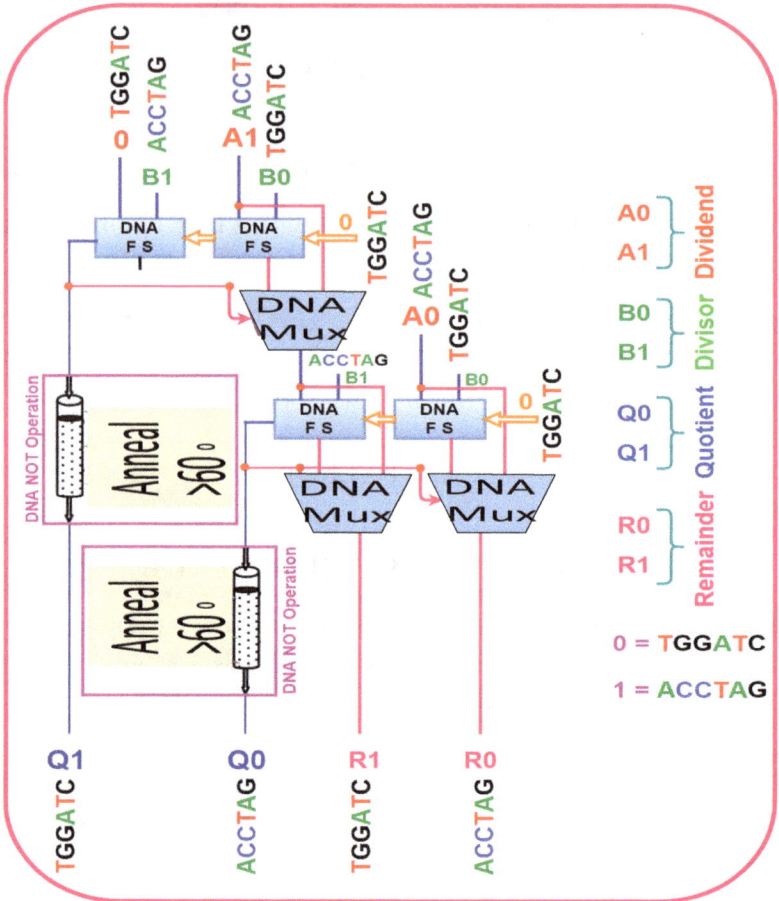

Figure 8.13.   DNA divider.

ACCTAG. Transistor switches are used to perform these operations as they can exist in only two possible states of a switch-open or closed. An open transistor is a device with no current, representing 0. A closed transistor, in which current passes through, means 1. Multiple transistors can be connected to accomplish the operations. The first transistor can be used to connect the second one and control the operation of the second transistor. The second transistor can be switched ON or OFF depending on the state of the second transistor. This DNA logical operation can allow or stop a current.

8.4.8 *Design procedure of DNA accumulator*

One DNA AND operation and two DNA D flip-flops are fed into the accumulator as input; four DNA NAND operations are performed to implement each DNA D-flip flop. The DNA AND operation have two inputs named LOAD and CLOCK. The output of the DNA AND operation will be the input of each DNA D flip-flop. And finally, performing all operations, each DNA D flip-flop evaluates the first and second outputs. Figure 8.14 shows the DNA accumulator circuit.

8.4.8.1 *Working principle of accumulator*

An accumulator is the special purpose register, a temporary memory location that stores an intermediate value in mathematical and logical operations. For example, if one wants to operate "4 + 2," the accumulator will hold the first value of 4 and then 6, the final deal. Let's consider the above example if the DNA AND operation's

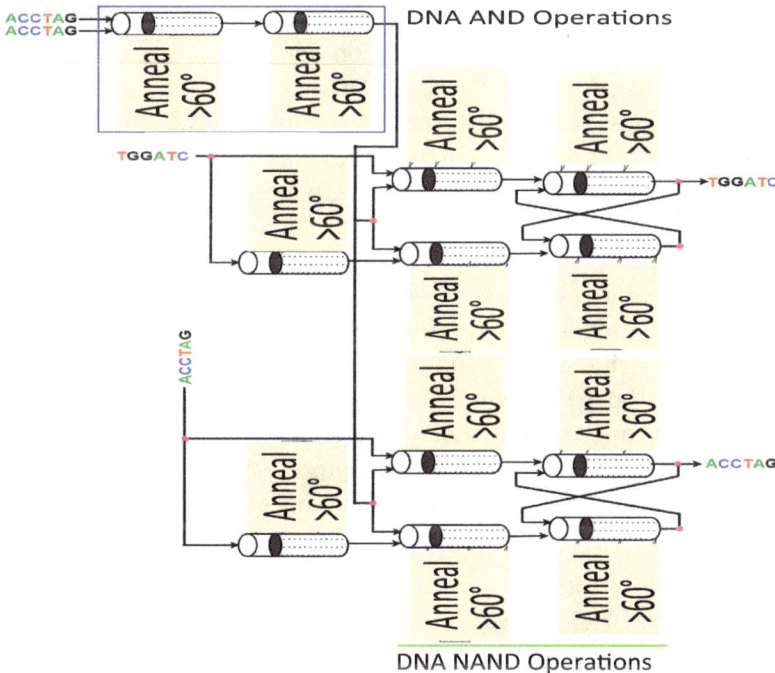

Figure 8.14. DNA accumulator.

output is ACCTAG or high value capable of storing the data into an accumulator. But if the output of AND operation is TGGATC or low, the accumulator will not store any data.

8.5 Applications

DNA computing is an emerging field of biomolecular computing, capable of performing arithmetic and logical operations that use DNA molecules as a carrier of information. One of the main advantages of DNA computing is miniaturization and parallelism over the traditional silicon-based machines. For example, a mix of 1018 strands of DNA could operate at 10,000 times the speed of today's advanced supercomputer (parker). Recently DNA computations have drawn researchers' attention. Researchers are paying attention to develop DNA computer or application problems. A few applications of DNA computing are security, cryptography, and forecasting, clustering, scheduling, encryption. An instruction detection model is the most recent development that has been built deploying DNA algorithms in cryptography. An instruction detection is a software application used to monitor a network or system's envious activity or policy violations. Researchers are proposing a DNA algorithm that will be possible to use in security technology. For instance, Adleman *et al.* and Boneh *et al.* proposed a model that can break a data encryption standard (DES). It is used for encrypting data technology.

Besides DNA cryptography and DES, DNA certification is another most mature development studied widely. Zhixing *et al.* proposed another algorithm to solve the job scheduling problem. He demonstrated the working operation of this model mimicking the Hamiltonian Path Problem. To solve the clustering problem, Bakar *et al.* have proposed a model. To assign edges and vertices DNA based clustering uses strands which helps to reduce time complexity because of having high parallelism features of DNA.

8.6 Summary

DNA computing is now one of the most exciting areas to be explored by researchers. There are many opportunities in expanding and manipulating DNA computing and operations to solve any real applications. It is possible to solve industrial engineering and management engineering problems using DNA NP-based computers. In this chapter, the DNA computing NP is explained, which integrates all the steps of DNA computing into one machine. The proposed DNA NP will solve any complex combinatorial problem. Besides that, it must ease solving the scheduling problem, clustering, and DES according to the logical instructions. For performing logical operations, DNA sequences ACCTAG and TGGATC are used. The test tube's base sequence is predetermined, which reacts chemically with given inputs. They need to provide temperature passed from the outside environment to accomplish chemical reactions. That is why the heat of each component is measured to observe how much temperature should deliver from the outdoor environment for performing any logical operation.

Bibliography

Max H. Garzon and Russell J. Deaton. Codeword design and information encoding in DNA ensembles. *Natural Computing*, 3(3): 253–292, 2004.

William H. Grover and Richard A. Mathies. An integrated microfluidic processor for single nucleotide polymorphism-based DNA computing. *Lab on a Chip*, 5(10): 1033–1040, 2005.

Jong Wook Hong, Vincent Studer, Giao Hang, W. French Anderson, and Stephen R. Quake. A nanoliter-scale nucleic acid processor with parallel architecture. *Nature Biotechnology*, 22(4): 435–439, 2004.

Peter M. Kogge, Toshio Sunaga, Hisatada Miyataka, Koji Kitamura, and Eric Retter. Combined DRAM and logic chip for massively parallel systems. In *Proceedings Sixteenth Conference on Advanced Research in VLSI*, pp. 4–16. IEEE, 1995.

Dominique Lavenier, Jean-Francois Roy, and David Furodet. DNA mapping using processor-in-memory architecture. In *2016 IEEE International Conference on Bioinformatics and Biomedicine (BIBM)*, pp. 1429–1435. IEEE, China, 2016.

Xingguo Liang, Hidenori Nishioka, Nobutaka Takenaka, and Hiroyuki Asanuma. A DNA nanomachine powered by light irradiation. *ChemBioChem*, 9(5): 702–705, 2008.

AB MacConnell. Price AK paegel BM ACS comb. An integrated microfluidic processor for DNA-encoded combinatorial library functional screening. *Science*, 19: 181–192, 2017.

Mayukh Sarkar, Prasun Ghosal, and Saraju P. Mohanty. Exploring the feasibility of a DNA computer: Design of an ALU using sticker-based DNA model. *IEEE Transactions on Nanobioscience*, 16(6): 383–399, 2017.

Ankur Sarker, Hafiz Md Hasan Babu, and Sarker Md Mahbubur Rashid. Design of a DNA-based reversible arithmetic and logic unit. *IET Nanobiotechnology*, 9(4): 226–238, 2015.

Haomiao Su, Jinglei Xu, Qi Wang, Fuan Wang, and Xiang Zhou. High-efficiency and integrable DNA arithmetic and logic system based on strand displacement synthesis. *Nature Communications*, 10(1): 1–8, 2019.

Theresa B. Taylor, Emily S. Winn-Deen, Enrico Picozza, Timothy M. Woudenberg, and Michael Albin. Optimization of the performance of the polymerase chain reaction in silicon-based microstructures. *Nucleic Acids Research*, 25(15): 3164–3168, 1997.

Yan-Feng Wang, Guang-Zhao Cui, Bu-Yi Huang, Lin-Qiang Pan, and Xun-Cai Zhang. DNA computing processor: An integrated scheme based on biochip technology for performing DNA computing. In *International Conference on Intelligent Computing*, pp. 248–257. Springer, China, 2006.

D.Y. Zhang and G. Seelig. Dynamic DNA nanotechnology using strand-displacement reactions. *Nature Chemical* 3(2): 103–113, 2011.

Chapter 9

Heat Calculation Techniques in DNA Computing

9.1 Introduction

Instead of using typical silicon chips, DNA computing uses biological molecules to do computations. The four-character genetic alphabets (A — adenine, G — guanine, C — cytosine, and T — thymine) are used in DNA computing instead of the binary alphabet (1 and 0) utilized by standard computers. This is possible due to the ability to create small DNA molecules with any arbitrary sequence. The input of any DNA operation can be represented by DNA molecules with specific sequences. The instructions are carried out by laboratory operations on the molecules, and the result is defined as some property of the final set of molecules. DNA computing promises significant and meaningful linkages between computers and life systems, as well as massively parallel computations. DNA computing can actually carry out millions of operations at the same time. Nowadays, required heat for DNA computing is an arresting matter for researchers. This chapter is going to express some ways to find out the amount of heat required for DNA computation.

9.2 Basic Definitions for Heat Calculation in DNA Computing

DNA has the characteristics of enabling classical logical operation using DNA sequence. DNA prefers to be in double-stranded form, while single-stranded DNA naturally migrates toward complementary sequences to form double-stranded complexes. Complementary sequences pair the bases adenine (A) with thymine (T) and cytosine (C) with guanine (G). DNA sequences pair in an antiparallel manner, with the 5' end of one sequence pairing with the 3' end of the complementary sequence.

Each input of the DNA operation will be the single standard sequence, if one is true, the complementary DNA sequence will be false. Suppose if ACTCGT is the input sequence then the complementary sequence will be TGAGCA. In DNA computing, when designing the DNA circuit, a predetermined single-strand sequence can be supplied to induce an appropriate chemical reaction. This sequence also helps to evaluate the output value whether it is true or false.

When the mixing step appears, it is needed to mix the two sequences to achieve a union of DNA sequences. In the mixing step, it is needed to give some heat to mix these. Then annealing appears and in annealing, it is needed to cool this little and make a double sequence bond. After annealing the step appears which is melting. In melting, need to heat the double-strand DNA sequence to make them a single strand complementary sequence and this sequence will be used in the DNA operations after some steps. So, the DNA melting temperature should be known:

1. **Nearest Neighbors**

   Depending on the nature of the sequence, one of two methods should be used to calculate melting temperature, T_m. Nearest Neighbors and Basic are the two methods that are discussed as follows:

   $$T_m = \frac{\Delta H}{A + \Delta S + R ln\left(\frac{C}{4}\right)} - 273.15 + 16.6 \log[\text{Na}^+], \qquad (9.1)$$

   where

   (a) Tm = melting temperature in °C,
   (b) ΔH = enthalpy change in kcal mol^{-1} (accounts for the energy change during annealing/melting),
   (c) A = constant of -0.0108 kcal K^{-1} mol^{-1} (accounts for helix initiation during annealing/melting),

(d) ΔS = entropy change in kcal K^{-1} mol^{-1} (accounts for energy unable to do work, i.e. disorder),

(e) R = gas constant of 0.00199 kcal K^{-1} mol^{-1} (constant that scales energy to temperature),

(f) C = oligonucleotide concentration in M or mol L^{-1} (use 0.0000005, i.e. 0.5 μM),

(g) -273.15 = conversion factor to change the expected temperature in Kelvins to °C,

(h) Na$^+$ = sodium ion concentration in M or mol L^{-1} (use 0.05, i.e. 50 mM).

This example will demonstrate the manual calculation of the Tm for the following sequence:

5'-AAAAACCCCCGGGGGTTTTT-3'

This is the above sequence paired with its reverse complement:

5'-AAAAACCCCCGGGGGTTTTT-3'

3'-TTTTTGGGGGCCCCCAAAAA-5'

$$T_m = \frac{\Delta H}{A + \Delta S + Rln\left(\frac{C}{4}\right)} - 273.15 + 16.6\log[\text{Na}^+]$$

$$T_m = \frac{-185.7 \text{ kcal mol}^{-1}}{\begin{array}{l} -0.0108 \text{ kcal K}^{-1} \cdot \text{mol}^{-1} + -0.4672 \text{ kcal K}^{-1} \cdot \text{mol}^{-1} \\ + 0.00199 \text{ kcal K}^{-1} \cdot \text{mol}^{-1} \times \ln \times \left(\frac{0.0000005 \text{ mol L}^{-1}}{4}\right) \\ - 273.15 + 16.6\log[0.05 \text{ mol L}^{-1}] \end{array}}$$

$T_m = 69.6$ °C.

2. **Basic Method**

A secondary method is used to calculate T_m is the basic method of a modified Marmur Doty formula, which is used for oligonucleotides with short sequences lengths (those that are 14 bases or less) [7, 8]. To calculate T_m the modified Marmur Doty formula is given as follows:

$$\boldsymbol{T_m = 2(A + T) + 4(C + G) - 7}$$

(a) T_m = melting temperature in °C,

(b) A = number of adenosine nucleotides in the sequence,

(c) T = number of thymidine nucleotides in the sequence,

(d) C = number of cytidine nucleotides in the sequence,

(e) G = number of guanosine nucleotides in the sequence,

(f) -7 = correction factor accounting for in solution.

So, for example, the melting temperature of a DNA sequence in different DNA operations can be calculated as DNA AND, DNA OR, DNA NOT, DNA NAND, DNA NOR, DNA XOR, and DNA XNOR.

1. **Heat Calculation of DNA AND Operation**

   Figure 9.1 shows the DNA AND operation. Here, ACCTAG = True and TGGATC= False. False and True inputs are given, then False output is obtained.

   Here, Input 1 = TGGATC

$$T_{m1} = 2(A + T) + 4(C + G) - 7$$
$$= 2(1 + 2) + 4(1 + 2) - 7$$
$$= 11.0\ °C.$$

Again, Input 2 = ACCTAG

$$T_{m2} = 2(A + T) + 4(C + G) - 7$$
$$= 2(1 + 1) + 4(2 + 1) - 7$$
$$T_{m2} = 9.0\ °C.$$

TGGATC + ACCTAG

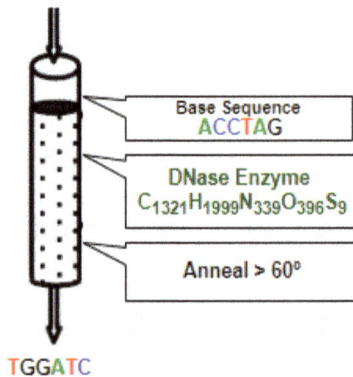

Figure 9.1.   DNA AND operation.

Other processes should also be done for finding an output in DNA computing. That's why another generalized process has to be performed within all steps and all the following steps are applicable for each tube for performing a DNA operation.

Preparing, Mixing, and Annealing: Allosteric DNAzyme-based DNA logic circuit, described a procedure to make a DNAzyme-based logic circuit. Here, All DNA logic operations were formed by annealing twice: first, the mixture of the inhibitor DNA strands and E6-type DNAzymes in $1\times$ TAE/Mg$_2$+ buffer (40 mM Tris, 20 mM acetic acid, 1 m MEDTA$_2$Na and 12.5 mM Mg(OAc)$_2$, pH 8.0) was heated at 95 °C for 4 min, 65 °C for 30 min, 50 °C for 30 min, 37 °C for 30 min, 22 °C for 30 min, and preserved at 20 °C; and then the substrates were added into the annealed mixture and incubated at constant temperature 20 °C for 4 h (total 6 h for preparing the DNA logic operation).

Melting, Amplifying, Separating, Extracting, Cutting, Ligating, Substituting, Marking, and Destroying Sequences: After that DNA logic circuits were triggered through displacement reaction in $1\times$ TAE/Mg2+ buffer (40 mM Tris, 20 mM acetic acid, 1 mM EDTA$_2$Na, and 12.5 mM Mg (OAc)$_2$, pH 8.0). The input DNA strands were added to a solution containing DNA logic circuits and reacted for >2 h at 20 °C. Next, the displaced products were stored at 20 °C for native PAGE or fluorescence detection. In addition, polyacrylamide gel electrophoresis (PAGE) needs 2 h and the PCR process for fluorescence detection needs less than 2 h.

Detecting and Reading Sequences: Here to describe a specific biochemical process briefly which serves as the basis of the DNA computing approach as polymerase chain reaction (PCR). Polymerases perform several functions, including the repair and duplication of DNA. PCR is a process that quickly amplifies the amount of specific DNA molecules in a given solution, using primer extension by the polymerase.

Each cycle of the reaction doubles the quantity of this molecule, leading to an exponential growth in the number of sequences. It consists of the following key processes:

(a) **Initialization:** A mixed solution of template, primer, dNTP
    and enzyme is heated to 94–98 °C for 1–9 min to ensure that
    most of the DNA template and primers are denatured;

(b) **Denaturation:** Heat the solution to 94–98 °C for 20–30 s for
    separation of DNA duplexes;

(c) **Annealing:** Lower the temperature enough (usually between
    50–64 °C) for 20–40 s for primers to anneal specifically to the
    ssDNA template;

(d) **Elongation/Extension:** Raise temperature to optimal elon-
    gation temperature of *Taq* or similar DNA polymerase
    (70–74 °C) for the polymerase adds dNTP's from the direction
    of 5′ to 3′ that are complementary to the template;

(e) **Final Elongation/Extension:** After the last cycle, a
    5–15 min elongation may be performed to ensure that any
    remaining ssDNA is fully extended.

Steps 2–4 are repeated 20–35 times; fewer cycles result in less
product, too many cycles increase the fraction of incomplete and
erroneous products. PCR is a routine job in the laboratory that
can be performed by an apparatus named thermal cycler. According to the PCR process, to produce an operational output of DNA
computation, it needs around 2 h.

The specific steps with heat for DNA computation are as
follows:

1.	DNA operation preparing	(98 °C–94 °C)
2.	Synthesizing	(98 °C–94 °C)
3.	Mixing	(95 °C–22 °C)
4.	Annealing	(70 °C–20 °C)
5.	Melting	(Depends on the sequence)
6.	Amplifying	
7.	Separating	
8.	Extracting	
9.	Cutting	
10.	Ligating	20 °C
11.	Substituting	
12.	Marking	
13.	Destroying	
14.	Detecting and reading	(98 °C–25 °C)

So, in DNA AND operation, the overall maximum required heat is

$$= (98 + 98 + 95 + 70 + 11 + 20 + 98) \, °C,$$
$$= 490 \, °C,$$

And the minimum required heat is $= (94 + 94 + 22 + 20 + 9 + 20 + 25) \, °C,$
$$= 284 \, °C.$$

Again, in DNA AND operation, all the processes are occurring in the test tube after mixing is completed. Here sometimes, needs to keep the temperature high, and sometimes it needs to keep the temperature low for several steps. So, need to keep the temperature at a maximum of 94 °C–98 °C. When DNA logic operation applies, it needs to keep the temperature around 20 °C and at detection time, it needs to keep it as 25 °C.

2. **Heat Calculation of DNA OR Operation**

   Figure 9.2 shows the DNA OR operation. Here, ACCTAG = True and TGGATC= False. False and True inputs are given, and then True output is obtained.

   Here, Input 1 = TGGATC

$$T_{m1} = 2(A + T) + 4(C + G) - 7$$
$$= 2(1 + 2) + 4(1 + 2) - 7$$
$$= 11.0 \, °C.$$

TGGATC + ACCTAG

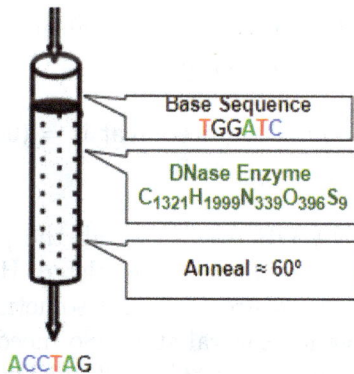

Base Sequence
TGGATC

DNase Enzyme
$C_{1321}H_{1999}N_{339}O_{396}S_9$

Anneal ≈ 60°

ACCTAG

Figure 9.2. DNA OR operation.

Again,

$$\text{Input } 2 = \text{ACCTAG}$$

$$= 2(A + T) + 4(C + G) - 7$$

$$= 2(1 + 1) + 4(2 + 1) - 7$$

$$T_{m2} = 9.0 \ °\text{C}.$$

Specific steps with heat for DNA computations are given as follows:

1.	DNA operation preparing	(98 °C–94 °C)
2.	Synthesizing	(98 °C–94 °C)
3.	Mixing	(95 °C–22 °C)
4.	Annealing	(70 °C–20 °C)
5.	Melting	(Depends on the sequence)
6.	Amplifying	
7.	Separating	
8.	Extracting	
9.	Cutting	
10.	Ligating	20 °C
11.	Substituting	
12.	Marking	
13.	Destroying	
14.	Detecting and reading	(98 °C–25 °C)

So, in DNA OR operation, the overall maximum required heat is
$= (98 + 98 + 95 + 70 + 11 + 20 + 98) \ °\text{C}$,
$= 490 \ °\text{C}$,

And the minimum required heat is $= (94 + 94 + 22 + 20 + 9 + 20 + 25) \ °\text{C}$,
$= 284 \ °\text{C}$

Again, in DNA OR operation, all the processes happening in the test tube after mixing are completed. Here sometimes it needs to keep the temperature high and sometimes needs to keep the temperature low for several steps. So, need to keep the temperature at a maximum of 94 °C–98 °C. When DNA logic operation applies then it needs to keep the temperature around 20 °C and at detection time it is as 25 °C.

TGGATC

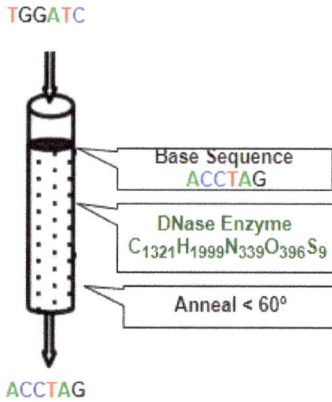

Figure 9.3.   DNA NOT operation.

3. Heat Calculation of DNA NOT Operation

Figure 9.3 shows the DNA NOT operation, where one input TGGATC (False) is given and the obtained output is ACCTAG (True).

Here, Input DNA sequence = TGGATC
So, melting temperature,

$$T_m = 2(A + T) + 4(C + G) - 7$$
$$= 2(1 + 2) + 4(1 + 2) - 7$$
$$= 11.0 \text{ °C}.$$

Specific steps with heat for DNA computations:

1.	DNA operation preparing	(98 °C–94 °C)
2.	Synthesizing	(98 °C–94 °C)
3.	Mixing	(95 °C–22 °C)
4.	Annealing	(70 °C–20 °C)
5.	Melting	(Depends on the sequence)
6.	Amplifying	
7.	Separating	
8.	Extracting	
9.	Cutting	
10.	Ligating	20 °C
11.	Substituting	
12.	Marking	
13.	Destroying	
14.	Detecting and reading	(98 °C–25 °C)

So, in DNA NOT operation, the overall maximum required heat
is

$$= (98 + 98 + 95 + 70 + 11 + 20 + 98) \text{ °C},$$
$$= 490 \text{ °C},$$

And the minimum required heat is $= (94 + 94 + 22 + 20 + 11 + 20 + 25)$ °C,

$$= 286 \text{ °C}.$$

Again, in DNA NOT operation, all the processes happening in
the test tube after mixing are completed. Here sometimes it needs
to keep the temperature high and sometimes it needs to keep the
temperature low for several steps. So, it is needed to keep the
temperature at a maximum of 94 °C–98 °C. When DNA logic
operation applies, it needs to keep the temperature around 20 °C
and at detection time, the temperature should be 25 °C.

4. **Heat Calculation of DNA XOR Operation**

   Figure 9.4 shows the DNA XOR operation and here also two
   inputs are given and one output is obtained by maintaining the
   truth table of DNA XOR operation.

   Again, Input$_1$ = ACCTAG

$$So, \ T_{m1} = 2(A + T) + 4(C + G) - 7$$
$$= 2(1 + 1) + 4(2 + 1) - 7$$
$$= 9.0°\text{C}.$$

ACCTAG + ACCTAG

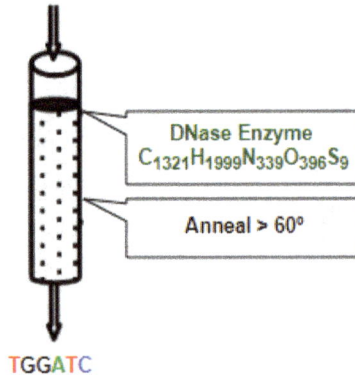

TGGATC

Figure 9.4.   DNA XOR operation.

Again, Input$_2$ = ACCTAG

$$So, \ T_{m2} = 2(A + T) + 4(C + G) - 7$$
$$= 2(1 + 1) + 4(2 + 1) - 7$$
$$= 9.0 \ °C.$$

Specific steps with heat for DNA computations are given below.

1.	DNA operation preparing	(98 °C–94 °C)
2.	Synthesizing	(98 °C–94 °C)
3.	Mixing	(95 °C–22 °C)
4.	Annealing	(70 °C–20 °C)
5.	Melting	(Depends on the sequence)
6.	Amplifying	
7.	Separating	
8.	Extracting	
9.	Cutting	
10.	Ligating	20 °C
11.	Substituting	
12.	Marking	
13.	Destroying	
14.	Detecting and reading	(98 °C–25 °C)

So, in DNA XOR operation, the overall maximum required heat is
= (98 + 98 + 95 + 70 + 11 + 20 + 98) °C,
= 490 °C,
 And minimum required heat = (94 + 94 + 22 + 20 + 9 + 20 + 25) °C,
= 284 °C.
 Again, in the DNA XOR operation, all the processes happening in the test tube after mixing are completed. Here in specific cases, it is needed to keep the temperature high and sometimes the temperature should be low for several steps. So, it is needed to keep the temperature at a maximum of 94 °C–98 °C. When DNA logic operation applies, it needs to keep the temperature around 20 °C and at detection time, the temperature is as 25 °C.

9.3 Heat Calculation in DNA Circuit

This section describes some DNA circuits to calculate the heat with an approximate value based on the theory discussed in Section 9.2.

9.3.1 *DNA full subtractor*

A full subtractor is a combinational circuit that performs subtraction of two DNA sequences, one is minuend and the other is subtrahend, taking into account the borrow of the previous adjacent lower minuend DNA sequence. This circuit has three inputs and two outputs. The three inputs A, B, and B_{in}, denote the minuend, subtrahend, and previous borrow, respectively. The two outputs, D and B_{out} represent the difference and output borrows, respectively. To create a DNA full subtractor, one DNA OR, two DNA AND, two DNA NOT, and two DNA XOR operations are required.

Here, three input sequence are as follows:

1. B_{in} = TGGATC,
2. A = ACCTAG,
3. B = ACCTAG.

Calculating the melting temperature of a specific DNA sequence:
For $Input_1$, B_{in} = TGGATC

$$\begin{aligned} T_{m(B_{in})} &= 2(A+T) + 4(C+G) - 7 \\ &= 2(1+2) + 4(1+2) - 7 \\ &= 11.0\ °C. \end{aligned}$$

Again, $Input_2$ and $Input_3$, A = B = ACCTAG

$$\begin{aligned} So,\ T_{m\ (A\ or\ B)} &= 2(A+T) + 4(C+G) - 7 \\ &= 2(2+1) + 4(2+1) - 7 \\ &= 11.0\ °C. \end{aligned}$$

Specific steps with heat for DNA full subtractor (for each tube) are given below.

1. DNA operation preparing   (98 °C–94 °C)
2. Synthesizing   (98 °C–94 °C)
3. Mixing   (95 °C–22 °C)
4. Annealing   (70 °C–20 °C)
5. Melting   (Depends on the sequence)
6. Amplifying
7. Separating
8. Extracting
9. Cutting
10. Ligating   20 °C
11. Substituting
12. Marking
13. Destroying
14. Detecting and reading   (98 °C–25 °C)

So, in DNA full subtractor, the overall maximum required heat is
$= (98 + 98 + 95 + 70 + 11 + 20 + 98)$ °C,
$= 490$ °C,

And the minimum required heat is $= (94 + 94 + 22 + 20 + 11 + 20 + 25)$ °C,
$= 286$ °C.

Again, in a specific basic DNA operations, all the processes happening in the test tube after mixing are completed. Here in specific cases, it is needed to keep the temperature high and sometimes the temperature should be low for several steps. So, the temperature should be kept at a maximum of 94 °C–98 °C. When the DNA logic operation applies, it needs to keep the temperature around 20°C and at detection time, it needs to keep as 25°C. Figure 9.5 shows the DNA circuit of the full subtractor.

9.3.2 *DNA full adder*

Full adder is the adder that adds three inputs and produces two outputs. The first two inputs are A and B and the third input is an input carry as C-IN. The output carry is designated as C_{out} and the normal output is designated as S which is SUM. A full adder logic is designed in such a manner that can take eight inputs together to create a byte-wide adder and cascade the carry DNA sequence from one adder to another. To create a DNA full adder, one DNA OR,

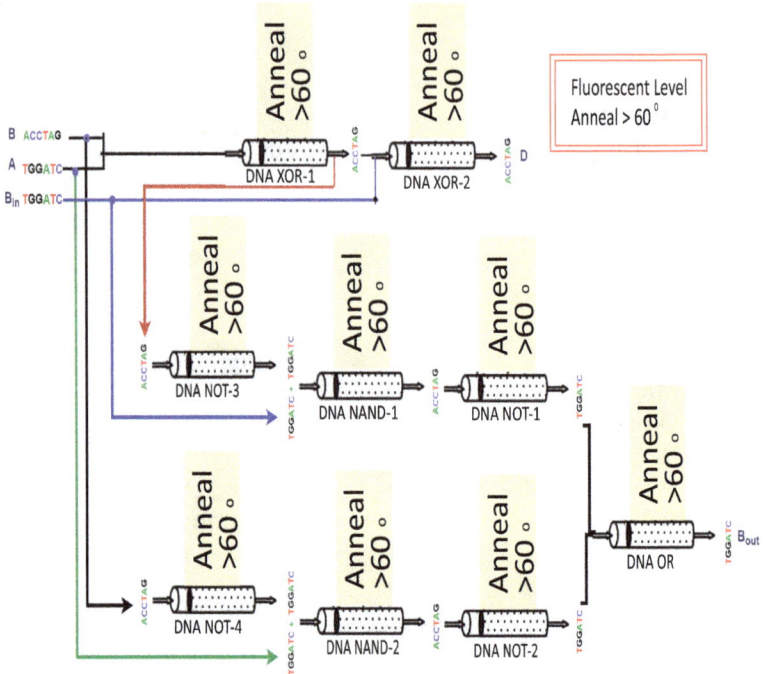

Figure 9.5. DNA full subtractor circuit.

two DNA NAND, one DNA XOR, and two DNA NOT operations are required. Figure 9.6 shows the DNA circuit of the full adder.

Here, three input DNA sequences are as follows:

1. A_0 = TGGATC,
2. A_1 = TGGATC,
3. A_2 = ACCTAG.

Calculating melting temperature of a specific DNA sequence:
For Input, A_0 and A_1 = TGGATC

$$T_{m\ (A0\ or\ A1)} = 2(A + T) + 4(C + G) - 7$$
$$= 2(1 + 2) + 4(1 + 2) - 7$$
$$= 11.0\ °C.$$

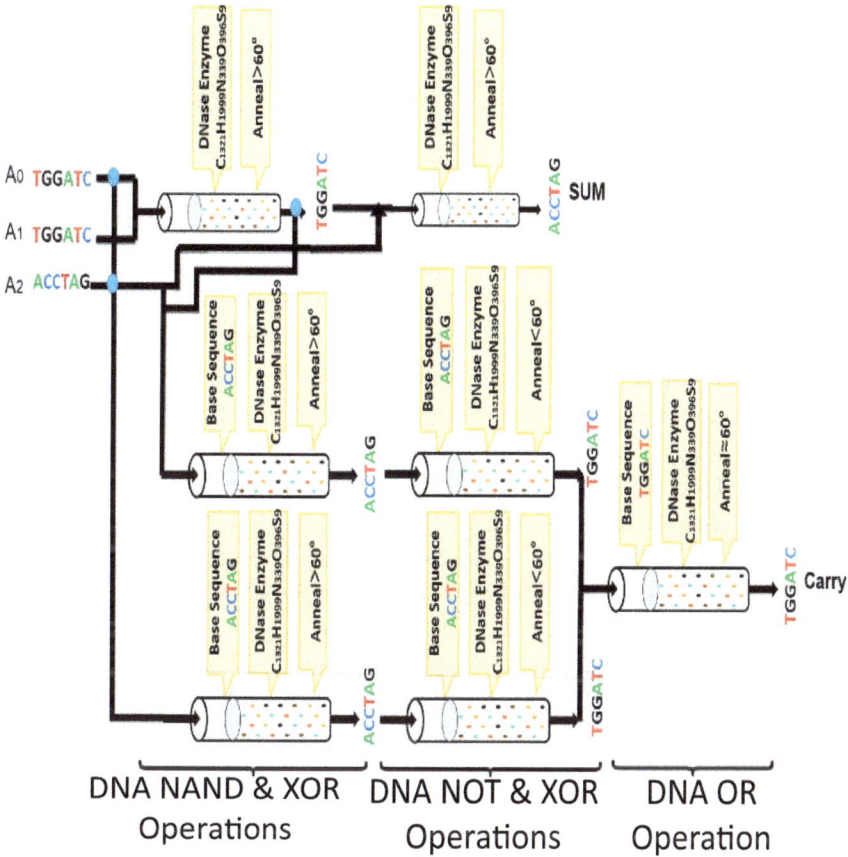

Figure 9.6. DNA full adder circuit.

Again, Input, $A_3 = \text{ACCTAG}$

$$So, \ T_{m \ (A3)} = 2(A + T) + 4(C + G) - 7$$
$$= 2(2 + 1) + 4(2 + 1) - 7$$
$$= 11.0 \ {}^\circ\text{C}.$$

Specific steps with heat for DNA full adder (for each tube):

1.	DNA operation preparing	(98 °C–94 °C)
2.	Synthesizing	(98 °C–94 °C)
3.	Mixing	(95 °C–22 °C)
4.	Annealing	(70 °C–20 °C)
5.	Melting	(Depends on the sequence)
6.	Amplifying	
7.	Separating	
8.	Extracting	
9.	Cutting	20 °C
10.	Ligating	
11.	Substituting	
12.	Marking	
13.	Destroying	
14.	Detecting and reading	(98 °C–25 °C)

So, in DNA full adder, the overall maximum required heat is
$= (98 + 98 + 95 + 70 + 11 + 20 + 98)$ °C,
$= 490$ °C.
And the minimum required heat is $= (94 + 94 + 22 + 20 + 11 + 20 + 25)$ °C,
$= 286$ °C.

Again, in a specific basic DNA operations, all the processes happening in the test tube after mixing are completed. Here in specific cases, it is needed to keep the temperature high and sometimes the temperature should be low for several steps. So, it is better to keep the temperature at a maximum of 94 °C–98 °C. When DNA logic operation applies, the temperature should be around 20 °C and at detection time, need to keep it as 25 °C.

9.3.3 *DNA multiplication circuit*

The multiplicand and multiplier can be of various numbers of DNA sequences. The product's DNA sequences depend on the number of DNA sequences of the multiplicand and multiplier. The number of DNA sequences of the product is equal to the sum of the number of DNA sequences of the multiplier multiplicand. To create a DNA Multiplication circuit, six DNA NAND, two DNA XOR, and

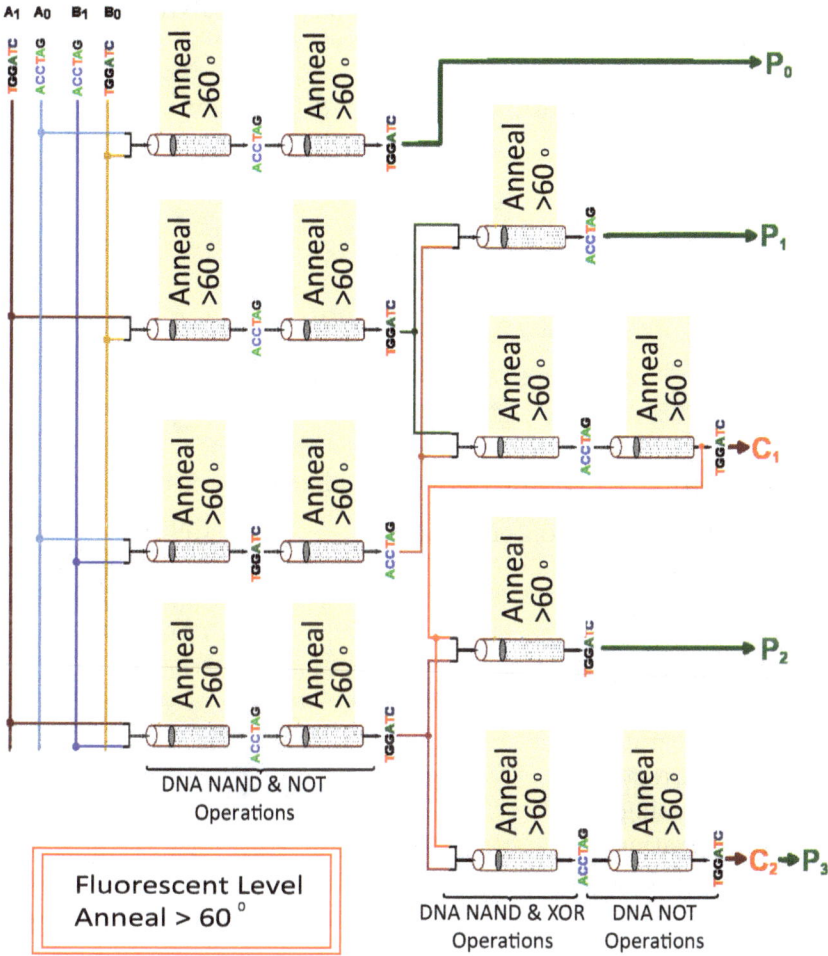

Figure 9.7. DNA multiplier circuit.

six DNA NOT operations are required. Figure 9.7 describes the DNA circuit of the 2-molecular multiplication.

Here, four input sequence are as follows:

1. A_0 = ACCTAG,
2. A_1 = TGGATC,
3. B_0 = TGGATC,
4. B_1 = ACCTAG.

Calculating the melting temperature of a specific DNA sequence:
For Input, A_0 and B_1 = ACCTAG

$$\begin{aligned}
T_{m \ (A0 \ or \ B1)} &= 2(A + T) + 4(C + G) - 7 \\
&= 2(2 + 1) + 4(2 + 1) - 7 \\
&= 11.0 \ ^\circ C.
\end{aligned}$$

Again, Input, A_1 and B_0 = TGGATC

$$\begin{aligned}
So, \ T_{m \ (A3)} &= 2(A + T) + 4(C + G) - 7 \\
&= 2(1 + 2) + 4(1 + 2) - 7 \\
&= 11.0 \ ^\circ C.
\end{aligned}$$

Specific steps with heat for DNA full adder (for each tube).

1.	DNA operation preparing	(98 °C–94 °C)
2.	Synthesizing	(98 °C–94 °C)
3.	Mixing	(95 °C–22 °C)
4.	Annealing	(70 °C–20 °C)
5.	Melting	(Depends on the sequence)
6.	Amplifying	
7.	Separating	
8.	Extracting	
9.	Cutting	
10.	Ligating	20 °C
11.	Substituting	
12.	Marking	
13.	Destroying	
14.	Detecting and reading	(98 °C–25 °C)

So, in DNA multiplier, the overall maximum required heat is
= (98 + 98 + 95 + 70 + 11 + 20 + 98) °C,
= 490 °C,
And the minimum required heat is = (94 + 94 + 22 + 20 + 11 + 20 + 25) °C,
= 286 °C.

Again, in a specific basic DNA operations, all the processes happening in the test tube after mixing are completed. Here in specific cases, the temperature should be kept high and sometimes low for several steps. So, the temperature must be at a maximum of

94 °C–98 °C. When DNA logic operation applies, it needs to keep the temperature around 20 °C, but at detection time, it needs to keep the temperature as 25 °C.

9.4   Applications

This section describes some real-life applications of DNA operation where it is used and provides superior performance against classical computing systems. In the case of different applications, classical computing systems might fail but DNA computing systems can show their capability.

Traveling salesman problem: The first theory of DNA computation was proposed by Leonard Adleman in 1994. He put his experimental theory to the test with a seven-point Hamiltonian path problem or also called the traveling salesman problem. The salesman in this problem needs to find the shortest path between seven cities whose distances are known in such a way that he crosses no city twice and returns to the original city. Adleman represented each city with a short DNA sequence with about 20 bases and a complementary strand with 20 bases as the street between the cities. All the fragments are capable of linking with each other. When the fragments were put in a tube and mixed, the natural bonding tendencies of the DNA created 109 formations or solutions in less than a second. Not all were correct and he took a week to extrapolate and filter out the shortest path through various techniques. Though this solution was not ideal, this demonstration opened floodgates to a wide range of possibilities and applications.

Security: Deploying DNA algorithms in cryptography to build an intrusion detection model is the most recent development. The ability to store 108 terabytes of data in 1 gram of DNA has led to the potential of holding a huge one-time pad. Another example is DNA steganography, in which a novel method was used to hide the messages in a microdot. Instead of the traditional binary encoding, each letter was denoted by three chemical bases, i.e. the letter A was encoded by CGA. These messages are then encoded into DNA sequences and concealed by mixing them in a tube with a large amount of sonicated random human DNA. This led to the formation of microdots, which were then decoded by the receiver with appropriate primers (short sequence with complementary bases). However, such encryption techniques have been posed with problems. The lack of a theoretical basis to explain the implementation and come up with good schemes seem to be a challenge. These are also expensive to apply, and analyze which requires modern infrastructure.

Artificial intelligence and machine learning: Artificial intelligence and machine learning are some of the prominent areas right now, as the emerging technologies have penetrated almost every aspect of humans' lives. However, as the number of applications increases, it becomes a challenging task for traditional computers to match up the accuracy and speed. And that's where DNA computing can help in processing complex problems in very little time, which would have taken traditional computers thousands of years.

9.5 Summary

DNA computing uses biological molecules to do computations. The four-character genetic alphabets (A — adenine, G — guanine, C — cytosine, and T — thymine) are used in DNA computing. The input of any DNA operation can be represented by DNA molecules with specific sequences. The instructions are carried out by laboratory operations on the molecules, and the result is defined as some property of the final set of molecules. DNA computing promises meaningful linkages between computers and life systems, as well as massively parallel computations. DNA computing can carry out millions of operations at the same time. Heat is an important property of any operation for computation. It is found that DNA computing needs heat in the test tube to execute the operation. Its different stage needs different amounts of heat, which is highlighted in this chapter.

Bibliography

Leonard M. Adleman. Molecular computation of solutions to combinatorial problems. *Science*, 266(5187): 1021–1024, 1994.

Diego Ricardo Alcoba, Roberto Carlos Bochicchio, Luis Lain, and Alicia Torre. On the measure of electron correlation and entanglement in quantum chemistry based on the cumulant of the second-order reduced density matrix. *The Journal of Chemical Physics*, 133(14): 144104, 2010.

Kenneth J. Breslauer, Ronald Frank, Helmut Blöcker, and Luis A. Marky. Predicting DNA duplex stability from the base sequence. *Proceedings of the National Academy of Sciences*, 83(11): 3746–3750, 1986.

Lajos Diósi. Qubit thermodynamics. In *A Short Course in Quantum Information Theory*, pp. 123–133. Springer, Singapore, 2011.

Susan M. Freier, Ryszard Kierzek, John A. Jaeger, Naoki Sugimoto, Marvin H. Caruthers, Thomas Neilson, and Douglas H. Turner. Improved free-energy parameters for predictions of RNA duplex stability. *Proceedings of the National Academy of Sciences*, 83(24): 9373–9377, 1986.

Luis A. Marky and Kenneth J. Breslauer. Calculating thermodynamic data for transitions of any molecularity from equilibrium melting curves. *Biopolymers: Original Research on Biomolecules*, 26(9): 1601–1620, 1987.

David H. Mathews, Jeffrey Sabina, Michael Zuker, and Douglas H. Turner. Expanded sequence dependence of thermodynamic parameters improves prediction of RNA secondary structure. *Journal of Molecular Biology*, 288(5): 911–940, 1999.

John SantaLucia Jr. A unified view of polymer, dumbbell, and oligonucleotide DNA nearest-neighbor thermodynamics. *Proceedings of the National Academy of Sciences*, 95(4): 1460–1465, 1998.

Junzo Watada. DNA computing and its application. In *Computational Intelligence: A Compendium*, pp. 1065–1089. Springer, Singapore, 2008.

Xuedong Zheng, Jing Yang, Changjun Zhou, Cheng Zhang, Qiang Zhang, and Xiaopeng Wei. Allosteric DNAzyme-based DNA logic circuit: Operations and dynamic analysis. *Nucleic Acids Research*, 47(3): 1097–1109, 2019.

Chapter 10

Speed Calculation Techniques in DNA Computing

10.1 Introduction

DNA computing is a kind of natural computing that uses the molecular characteristics of DNA to conduct logical and arithmetic operations instead of typical carbon/silicon chips. The four-character genetic alphabets (A — adenine, G — guanine, C — cytosine, and T — thymine) are used in DNA computing instead of the binary digits (1 and 0) utilized by standard computers. This enables massively parallel computation, making it possible to answer difficult mathematical equations or problems in a fraction of the time. As a result, computation is far more efficient with a large volume of self-replicating DNA than with a standard computer, which would require a lot more hardware. Information or data will now be kept in the form of the bases A, T, G, and C, rather than binary digits. The capacity to generate short DNA sequences artificially allows these sequences to be used as inputs for algorithms. This is possible due to the ability to create small DNA molecules with any arbitrary sequence. The input of any DNA operation can be represented by DNA molecules with specific sequences. The instructions are carried out by laboratory operations on the molecules, and the result is defined as some property of the final set of molecules. DNA computing promises significant and meaningful linkages between computers and life systems, as well as massively parallel computations.

DNA computing can actually carry out millions of operations at the same time.

Besides the accuracy of a system, time or speed is a metric that can be used to measure the performance of a system. This chapter describes how to calculate speed or consume time for different operations in DNA computation systems.

10.2   Speed Calculation in DNA Computing

DNA molecules can be used as information storage media. Usually, DNA sequences of around 8–20 base pairs are used to represent DNA sequences, and numerous methods have been developed to manipulate and evaluate them. In order to manipulate a wet technology to perform computations, one or more of the following techniques are used as computational operators for copying, sorting, splitting, or concatenating the information contained within DNA molecules as ligation, hybridization, polymerase chain reaction (PCR), gel electrophoresis, and enzyme reaction.

Allosteric DNAzyme-based DNA logic circuit, described a procedure to make a DNAzyme-based logic circuit. Here, All DNA logic operations were formed by annealing twice: firstly, the mixture of the inhibitor DNA strands and E6-type DNAzymes in 1× TAE/Mg2+ buffer (40 mM Tris, 20 mM acetic acid, 1 m MEDTA$_2$Na and 12.5 mM Mg(OAc)$_2$, pH 8.0) was heated at 95 °C for 4 min, 65 °C for 30 min, 50 °C for 30 min, 37 °C for 30 min, 22 °C for 30 min, and preserved at 20 °C; and then the substrates were added into the annealed mixture and incubated at constant temperature 20 °C for 4 h (total 6 h for preparing the DNA logic operation).

After that Logic operations were triggered through displacement reaction in 1× TAE/Mg2+ buffer (40 mM Tris, 20 mM acetic acid, 1 mM EDTA$_2$Na, and 12.5 mM Mg (OAc)$_2$, pH 8.0). The input DNA strands were added to a solution containing DNA logic operations and reacted for > 2 h at 20 °C. Next, the displaced products were stored at 20 °C for native PAGE or fluorescence detection. In addition, polyacrylamide gel electrophoresis (PAGE) needs 2 h and the PCR process for fluorescence detection needs less than 2 h.

A specific biochemical process described briefly serves as the basis of the DNA computing approach as PCR. Polymerases perform several functions, including the repair and duplication of DNA. PCR is a process that quickly amplifies the amount of specific DNA molecules in a given solution, using primer extension by the polymerase. Each cycle of the reaction doubles the quantity of this molecule, leading to an exponential growth in the number of sequences. It consists of the following key processes:

1. **Initialization:** A mixed solution of template, primer, dNTP, and the enzyme is heated to 94 °C–98 °C for 1–9 min to ensure that most of the DNA template and primers are denatured;
2. **Denaturation:** Heat the solution to 94 °C–98 °C for 20–30 s for separation of DNA duplexes;
3. **Annealing:** Lower the temperature enough (usually between 50 °C and 64 °C) for 20–40 s for primers to anneal specifically to the ssDNA template;
4. **Elongation/Extension:** Raise temperature to optimal elongation temperature of *Taq* or similar DNA polymerase (70 °C–74 °C) for the polymerase adds dNTP's from the direction of $5'$ to $3'$ that are complementary to the template;
5. **Final Elongation/Extension:** After the last cycle, a 5–15 min elongation may be performed to ensure that any remaining ssDNA is fully extended.

Steps 2–4 are repeated 20–35 times; fewer cycles results in less product, too many cycles increase the fraction of incomplete and erroneous products. PCR is a routine job in the laboratory that can be performed by an apparatus named a thermal cycler. According to the PCR process, to produce an operational output of DNA computation, it needs around 2 h.

So, except for initial preparation and the last phase of fluorescence detection for each operation in a particular test tube, it needs more or less 2 h.

10.3   Speed Calculation in DNA Circuits

This subsection is going to describe some DNA circuits to calculate their performing time or speed in an approximate value based on the theory described in Section 10.2.

10.3.1 *DNA full subtractor*

A full subtractor is a combinational circuit that performs 2-molecular subtraction, one is minuend and the other is subtrahend, taking into account the borrow of the previous adjacent lower minuend DNA sequence. This circuit has three inputs and two outputs. The three inputs A, B, and B_{in}, denote the minuend, subtrahend, and previous borrow, respectively. The two outputs, D and B_{out} represent the difference and output borrows, respectively. To create a DNA full subtractor, one DNA OR, two DNA AND, two DNA NOT, and two DNA XOR operations are required. Figure 10.1 describes the DNA circuit of a full subtractor. Here, two input will be DNA sequences and it also

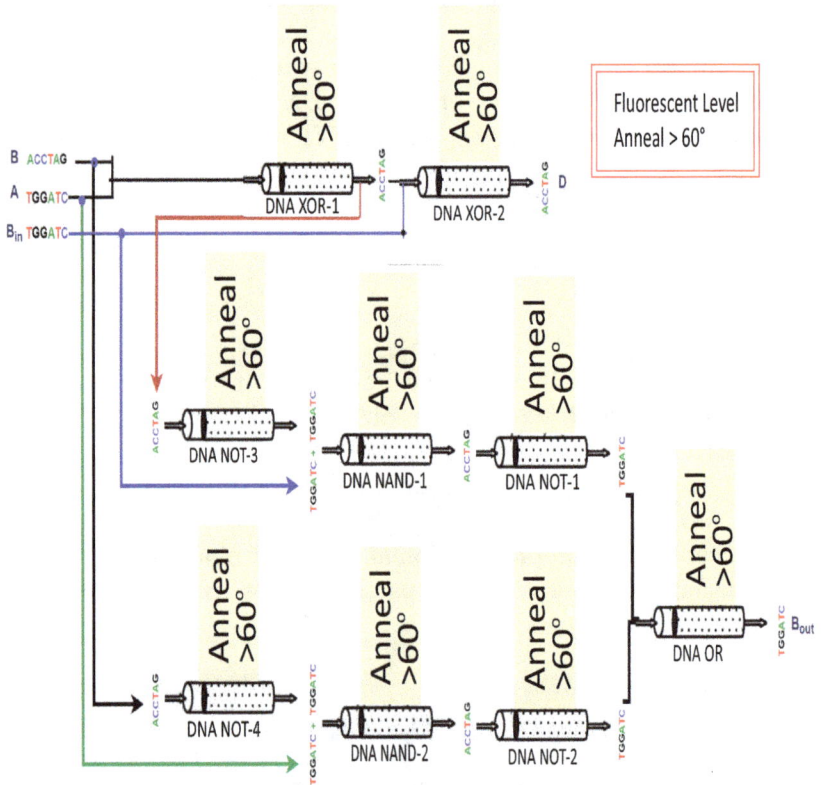

Figure 10.1.   DNA full subtractor circuit.

provides two output that contains DNA sequence. In DNA computing, DNA AND operation will be easy to represent by DNA NOT and DNA NAND operation.

According to Section 10.2, it is perceived that any DNA basic (i.e. AND, OR, NOT, and XOR) operation needs more or less 2 h to perform. In addition, 6 h are needed for preparing any DNA basic operation and 2 h for fluorescence detection which is fixed for any multi-(basic) basic DNA operation.

As the second pipeline is the largest pipeline for processing input to the output of the DNA full subtractor, it will be taken for measuring the total required performing time. Other basic DNA operations will be performed within this time in parallel.

So, required time for four basic operations in DNA is
$$= (2 + 2 + 2 + 2)$$
$$= 8 \text{ h};$$

where the performing time for DNA (AND, OR, NOT, and XOR) operation needs more or less 2 h.

The total performing time/required time for performing DNA full subtractor is the summation of initial preparation time, fluorescence detection time, and DNA operation time.

So, the required time for DNA multiplexer is
$$= (\text{basic DNA operation preparation time} + \text{four basic DNA}$$
$$\text{operations time} + \text{fluorescence detection})$$
$$= (6 + 8 + 2)$$
$$= 16 \text{ h (approximately)}.$$

10.3.2   *DNA full adder*

Full adder is the adder that adds three inputs and produces two outputs. The first two inputs are A and B and the third input is an input carry as C_{in}. The output carry is designated as C_{out} and the normal output is designated as S which is SUM. A full adder logic is designed in such a manner that can take eight inputs together to create a byte-wide adder and cascade the carry DNA sequence from one adder to another. To create a DNA full adder, one DNA OR, two DNA NAND, one DNA XOR, and two DNA NOT operations are required. Figure 10.2 depicts the DNA circuit of the full adder.

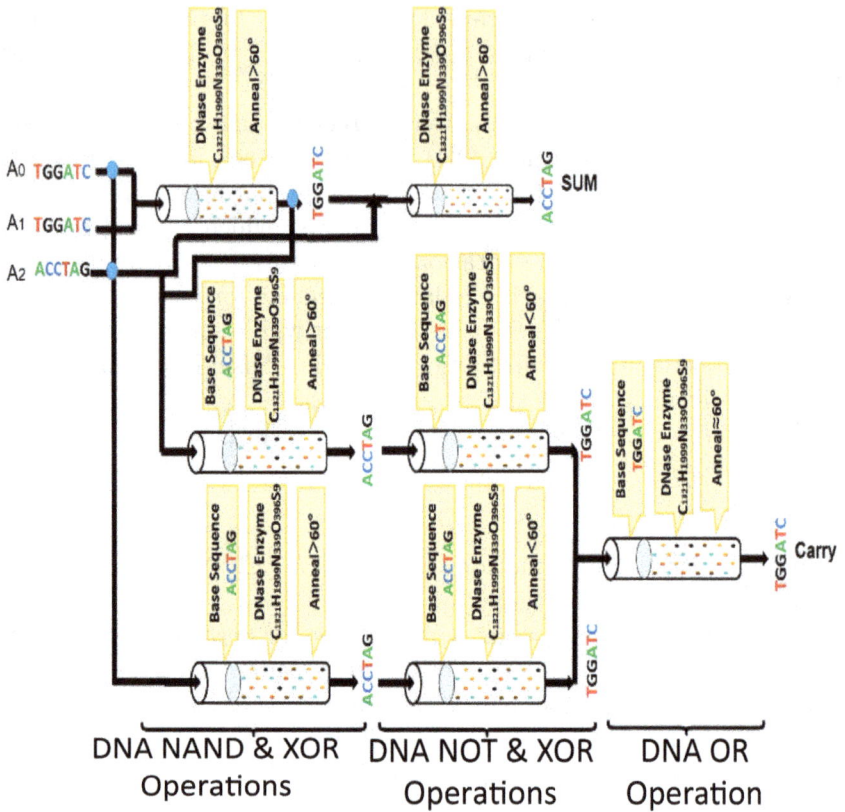

Figure 10.2.   DNA full adder circuit.

To find the required performing time of DNA full adder, it is divided into three pipelines as some of the basic DNA operations are performed in parallel. Three pipelines are as follows:

1. DNA XOR, DNAXOR;
2. DNA XOR, DNA NAND, DNA NOT, DNA OR;
3. DNA NAND, DNA NOT, DNA OR.

According to Section 10.2, it can be perceived that any DNA basic (i.e. AND, OR, NOT, and XOR) operation needs more or less 2 h to perform. In addition, 6 h are needed for preparing any DNA basic operation and 2 h for fluorescence detection which is fixed for any multi-(basic) DNA operation.

As the second pipeline is the largest pipeline for processing input to the output of the full adder, it is taken for measuring the total required performing time. Other DNA basic operations will be performed within this time in parallel.

So, the required time for four basic operations in DNA is
$$= (2 + 2 + 2 + 2)$$
$$= 8 \text{ h}.$$

where performing time for DNA (NAND, OR, NOT, and XOR) operation needs more or less 2 h.

The total performing time/required time for performing DNA full adder is a summation of initial preparation time, fluorescence detection time, and DNA operational time.

So, the required time for DNA full adder is
$$= (\text{basic DNA operation preparation time} + \text{four basic DNA}$$
$$\quad \text{operations time} + \text{fluorescence detection})$$
$$= (6 + 8 + 2) \text{ h}$$
$$= 16 \text{ h (approximately)}.$$

10.3.3 *DNA multiplication circuit*

A binary multiplier is a combinational logic circuit or digital device used for multiplying two binary numbers. The two numbers are more specifically known as multiplicand and multiplier and the result is known as a product. The multiplicand and multiplier can be of various numbers of DNA sequences. The product's DNA sequence number depends on the number of DNA sequences of the multiplicand and multiplier. The number of DNA sequences of the product is equal to the sum of the number of DNA sequences of the multiplier multiplicand. To create a DNA multiplication circuit, six DNA NAND, two DNA XOR, and six DNA NOT operations are required. Figure 10.3 describes the DNA circuit of the 2-molecular multiplication.

To find the required performing time of a DNA multiplication circuit, it is divided into multiple pipelines as some of the basic DNA operations are performed in parallel. Among all of the pipelines, five are as follows:

1. DNA NAND, DNA NOT;
2. DNA NAND, DNA NOT, DNA XOR;

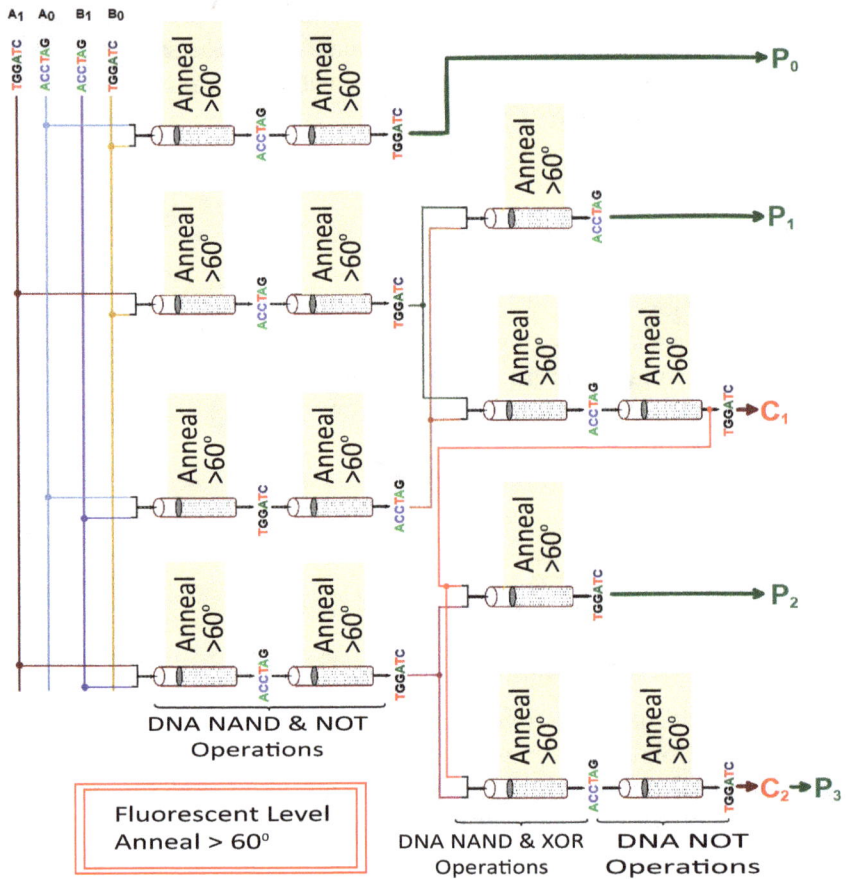

Figure 10.3.   DNA multiplier circuit.

3. DNA NAND, DNA NOT, DNA NAND, DNA NOT, DNA NAND,
   DNA NOT;
4. DNA NAND, DNA NOT, DNA NAND, DNA NOT;
5. DNA NAND, DNA NOT, DNA XOR.

According to Section 10.3, it can be perceived that any DNA basic
(i.e. AND, OR, NOT, and XOR) operation needs more or less 2 h
to perform its action. In addition, 6 h are needed for preparing any
DNA basic operation and 2 h for fluorescence detection which is fixed
for any DNA operation.

As the third pipeline is the longest pipeline for processing input to the output of the 2-molecular multiplication circuit, it is taken for measuring the total required performing time. Other DNA basic operations will be performed within this time in parallel.

So, required time for four basic operations in DNA is $= (2 \times 6) = 12$ h, where the required time for DNA (NAND and NOT) operations need more or less 2 h.

The total performing time/the required time for performing a 2-molecular DNA multiplication circuit is the summation of the initial preparation time, fluorescence detection time, and DNA operational time.

So, the required time for a 2-molecular DNA multiplication circuit is

$=$ (basic DNA operations preparation time $+$ six basic DNA operations time $+$ fluorescence detection)

$= (6 + 12 + 2)$ h

$= 20$ h (approximately).

10.4 Applications

This section describes some real-life applications of DNA operation, where it is used and provides superior performance against classical computing systems. In the case of different applications, classical computing systems might fail but DNA computing systems can show their capability.

Encryption: The data encryption standard (DES) employs a 56 DNA sequences key to encrypt 64 DNA sequence information. Breaking DES means finding a key that maps the plain-text to the cipher-text given a (plain-text, cipher-text) pair. A traditional DES assault would need an exhaustive search of all 256 DES keys, which would take 10,000 years at a rate of 100,000 operations per second. As a result, molecular programs were developed, which needed around 4 months of laboratory work instead.

DNA computation is expected to provide significant benefits in terms of speed, energy efficiency, and cost-effective information storage. The number of operations per second in Adleman's model might be as high as 1.2×10^{18}. This is around 1,200,000 times quicker

than the most powerful supercomputer. Additionally, DNA computers have the potential to be extremely energy efficient. In theory,
one joule is enough to do about 2×1019 ligation operations. This is
surprising given that the second law of thermodynamics states that
there may only be 34×10^{19} (irreversible) operations per joule in
theory (at room temperature). Existing supercomputers are significantly less efficient, with 10^9 operations per joule at most. Finally,
storing information in DNA molecules could allow for an information
density of around 1 DNA sequence per cubic nanometer, compared
to 1 DNA sequence per 10^{12} nm^3 in current storage media.

Weather Forecasting: Weather affects over 30% of the US GDP
($6 trillion) directly or indirectly, affecting food production, trans-
portation, and retail trade, among other things. The capacity to better predict the weather would be extremely beneficial in a variety
of sectors, not to mention giving people more time to prepare for
disasters. While scientists have long desire to do this, the equations
controlling such processes involve a large number of variables, making traditional simulation time-consuming. "Using a classical computer to undertake such analysis might take longer than it takes
for the actual weather to evolve!" said DNA researcher Seth Lloyd.
This prompted Lloyd and colleagues at MIT to demonstrate that the
weather equations have a hidden wave nature that can be solved by
a DNA computer. DNA computers could aid in the development of
improved climate models, allowing us to gain a better understanding
of how humans affect the ecosystem. These models form the foundation for the projections of future warming, and they assist us in
determining what steps need to be taken now to avoid calamities.

10.5   Summary

Speed calculation is an important part in DNA computing. DNA
computing process is another way of computing where the storage
capacity is huge and can work in a parallel way. This chapter has
presented the way to calculate the speed and time of DNA computing
in detail. Some examples have been presented to clear the concept of
speed calculation. Necessary figures and explanations are also shown
in this chapter.

Bibliography

Leonard M. Adleman. Molecular computation of solutions to combinatorial problems. *Science*, 266(5187): 1021–1024, 1994.

Nemanja Isailovic, Yatish Patel, Mark Whitney, and John Kubiatowicz. Interconnection networks for scalable quantum computers. In *33rd International Symposium on Computer Architecture (ISCA'06)*, pp. 366–377. IEEE, USA, 2006.

Lev B. Levitin, Tommaso Toffoli, and Zachary Walton. Operation time of quantum gates. arXiv preprint quant-ph/0210076, 2002.

David H. Mathews, Jeffrey Sabina, Michael Zuker, and Douglas H. Turner. Expanded sequence dependence of thermodynamic parameters improves prediction of RNA secondary structure. *Journal of Molecular Biology*, 288(5): 911–940, 1999.

Tzvetan S. Metodi and Frederic T. Chong. Quantum computing for computer architects. *Synthesis Lectures in Computer Architecture*, 1(1): 1–154, 2006.

Tzvetan S. Metodi, Darshan D. Thaker, Andrew W. Cross, Frederic T. Chong, and Isaac L. Chuang. A quantum logic array microarchitecture: Scalable quantum data movement and computation. In *38th Annual IEEE/ACM International Symposium on Microarchitecture (MICRO'05)*, 12 pp. IEEE, Spain, 2005.

C. Monroe, R. Raussendorf, A. Ruthven, K.R. Brown, P. Maunz, L.-M. Duan, and J. Kim. Large-scale modular quantum-computer architecture with atomic memory and photonic interconnects. *Physical Review A*, 89(2): 022317, 2014.

John SantaLucia Jr. A unified view of polymer, dumbbell, and oligonucleotide DNA nearest-neighbor thermodynamics. *Proceedings of the National Academy of Sciences*, 95(4): 1460–1465, 1998.

Darshan D. Thaker, Tzvetan S. Metodi, Andrew W. Cross, Isaac L. Chuang, and Frederic T. Chong. Quantum memory hierarchies: Efficient designs to match available parallelism in quantum computing. In *33rd International Symposium on Computer Architecture (ISCA'06)*, pp. 378–390. IEEE, USA, 2006.

Mark Whitney, Nemanja Isailovic, Yatish Patel, and John Kubiatowicz. Automated generation of layout and control for quantum circuits. In *Proceedings of the 4th International Conference on Computing Frontiers*, Italy, pp. 83–94, 2007.

Xuedong Zheng, Jing Yang, Changjun Zhou, Cheng Zhang, Qiang Zhang, and Xiaopeng Wei. Allosteric DNAzyme-based DNA logic circuit: Operations and dynamic analysis. *Nucleic Acids Research*, 47(3): 1097–1109, 2019.

Concluding Remarks

DNA computing is an area of natural computing based on the concept of performing logical and arithmetic operations using molecular properties of DNA by replacing traditional carbon/silicon chips with biochips. This allows massively parallel computation, where complex mathematical equations or problems can be solved in much less time. Hence with a considerable amount of self-replicating DNA, computation is much efficient than the traditional computer which would require a lot more hardware. A good experience with biology and computer science is required to build algorithms to be executed in DNA computing. Every single cell that makes up a living thing has information for numerous tasks required for the cell to survive. The nucleic acid molecules found in each cell are where the genetic material is kept. The term "deoxyribonucleic acid" (DNA) refers to the nucleic acid that is the most stable. Millions of linked nucleotides form the lengthy polymers known as helical structures on each of the DNA strands. These nucleotides are made up of a phosphate group, one of four nitrogen bases, and a five-carbon sugar. Adenine (A), Thymine (T), Guanine (G), and Cytosine (C) are nitrogen bases that encode genetic information, while the others offer structural stability. The base-pairing rule, which pairs T with A and C with G, connects the strands together. DNA computing is unique among traditional computer systems because of its parallel processing, vast storage capacity, and ability to do nano-level computing. In DNA computing, DNA logic operations offer unique properties such as stability and re-usability. DNA computing is a parallel computing. *DNA Logic Design: Computing with DNA* is a unique book where all basic and operational computations of binary logic design are shown with detailed explanations and appropriate figures. This book discusses the arithmetic operations in DNA computing. The combinational circuits and Memory devices are also described here with their working principles. It also discusses the circuit architecture and applications of DNA programmable logic devices and DNA nano processors. At the end of the book, there are two more chapters to explain the heat calculation techniques and speed calculation techniques. This is a complete book about DNA basics and DNA logic designs. The language used here to explain everything is so easy to give a clear concept of DNA logic design to the readers.

Index

www.ingramcontent.com/pod-product-compliance
Lightning Source LLC
Chambersburg PA
CBHW050546190326
41458CB00007B/1935